Hydrogen-bonded Capsules
Molecular Behavior in Small Spaces

Hydrogen-bonded Capsules
Molecular Behavior in Small Spaces

Julius Rebek, Jr.

Fudan University, China & The Scripps Research Institute, USA

World Scientific

NEW JERSEY · LONDON · SINGAPORE · BEIJING · SHANGHAI · HONG KONG · TAIPEI · CHENNAI · TOKYO

Published by

World Scientific Publishing Co. Pte. Ltd.
5 Toh Tuck Link, Singapore 596224
USA office: 27 Warren Street, Suite 401-402, Hackensack, NJ 07601
UK office: 57 Shelton Street, Covent Garden, London WC2H 9HE

Library of Congress Cataloging-in-Publication Data
Rebek, Julius, Jr.
  Hydrogen-bonded capsules : molecular behavior in small spaces / Julius Rebek, Jr.,
Fudan University, China & The Scripps Research Institute, USA.
    pages cm
  Includes bibliographical references and index.
  ISBN 978-9814678353 (hardcover : alk. paper) -- ISBN 981467835X (hardcover : alk. paper)
  1. Nanochemistry. 2. Microencapsulation. 3. Hydrogen bonding. 4. Chemistry, Organic. I. Title.
  QC176.8.N35R42 2015
  541'.2--dc23

                                        2015023647

British Library Cataloguing-in-Publication Data
A catalogue record for this book is available from the British Library.

For

Stephen L. Buchwald and Jin-Quan Yu

Architects of chemical catalysis

Contents

Prologue

Research on molecular recognition became a fashionable topic for physical organic chemists in the 1980s. At the time there was growing disenchantment with the nonclassical ion controversy, and orbital symmetry fatigue was taking hold in the community. But the discovery of crown ethers by Pedersen in 1962 ignited the study of their complexes with ions, and macrocyclic polyethers dominated the scene for decades. The weak intermolecular interactions of these complexes gave rise to the concepts of molecular recognition: complementarity of sizes, shapes and chemical surfaces. The culmination of this activity was the award of the Nobel Prize in Chemistry to Pedersen, Cram and Lehn in 1987 for establishing host–guest chemistry and, on a grander scale, supramolecular chemistry.

While most practitioners favored macrocyclic compounds for the recognition of ions, there were several groups who pursued receptors of different shapes for other molecular targets. The shapes included a number of concave surfaces and — in our case — clefts, optimized for the study of reversible interactions. Along with many others, we became proficient in sculpting molecular surfaces to complement the convex surfaces of the small molecular targets. This activity was a departure from targeting spherical ions with circular receptors — the low-hanging fruit — since neutral molecule targets come in all manner of sizes and shapes. The application of hydrogen bonding for recognition purposes offered much appeal and became suspiciously close to the base-pairing of DNA. Purines, pyrimidines and other heterocyclic bases, rich in hydrogen-bonding possibilities, became popular targets. We pursued synthetic receptors that contacted incresingly large fractions of surfaces of the target molecules, and it became a reasonable goal to engineer a system that completely surrounded the (convex) target molecule by the (concave) synthetic receptor. After all, enzymes and other biological receptors are capable of doing just that: folding around their targets, isolating them from the medium, surrounding them with a hydrophobic environment and presenting them with functional groups for signaling and catalysis. The present synthetic constructs are similar; interactions between the container and contained molecules are constant and encounters are not left to chance — they are prearranged, private and prolonged.

At that time we collaborated with Javier de Mendoza, and our most advanced receptor chelated adenosine phosphate, in such a way that most of the target molecule's surface was in contact with the synthetic receptor. We discussed the possibility of completely surrounding and isolating a target molecule — encapsulating it — through the assembly of self-complementary receptor pieces. This notion eventually became the "tennis ball," and started our journey on the road to hydrogen-bonded capsules and their open-ended cousins, the cavitands. Only recently have we been able to combine the two types of containers in that special solvent — water. Now, a little more than 20 years after we started, we explore our experiences with capsules in this monograph. Despite admonishments to the contrary, we will strive to tell the story chronologically, as far as a given container is concerned. We will also emphasize how we developed methods and interpreted our data at the time, and we make apologies in advance to other researchers who may feel left out. After all, this is a narrative of our own work.

CHAPTER 1

Spherical and Similar Capsules

A Drop Shot with a Tennis Ball

The ungainly shapes of nucleotides require unsymmetrical complements,[1] but for structures that are intended to completely surround a smaller, convex molecule, a sphere is not a bad starting point. The initial design[2] is shown in Figure 1.1; the compound **1** was intended to dimerize on account of its self-complementarity. The hydrogen bond donors of the glycolurils at the ends of the structure were to find their complements in the carbonyl acceptors of the imides at the center. The roughly spherical dimer **1.1** evokes a baseball or tennis ball, with the hydrogen bonds acting as the seams that hold it together. Rene Wyler set out to synthesize the compound but made an ingenious abbreviation of the original design. He realized that the *glycoluril carbonyls* could also act as acceptors in a self-complementary sense if a shorter spacer separated the molecule's ends. He hit a drop shot and produced the appropriate module **2** in a single chemical step from durene tetrabromide and the notoriously insoluble diphenyl glycoluril shown.[3] The condensation occurred in hot DMSO with KOH as the base, a recipe that the reader will recognize will also lead to polymerization, and it did. Fortunately, however, the polymer is insoluble while the module remains in solution. The yield was low (20%) but the isolation was easy. Much later, a variant of the procedure suitable for undergraduate laboratories was developed by Fraser Hof and Liam Palmer.[4] However, in 1993, we still intended to go for the baseline shot and synthesize the longer **1.1**. So we promoted its stature to the larger "softball" and referred to the smaller **2a.2a** as the "tennis ball."

The compound showed the expected NMR earmarks of a hydrogen-bonded dimer **2a.2a**, but it was not until a guest could be

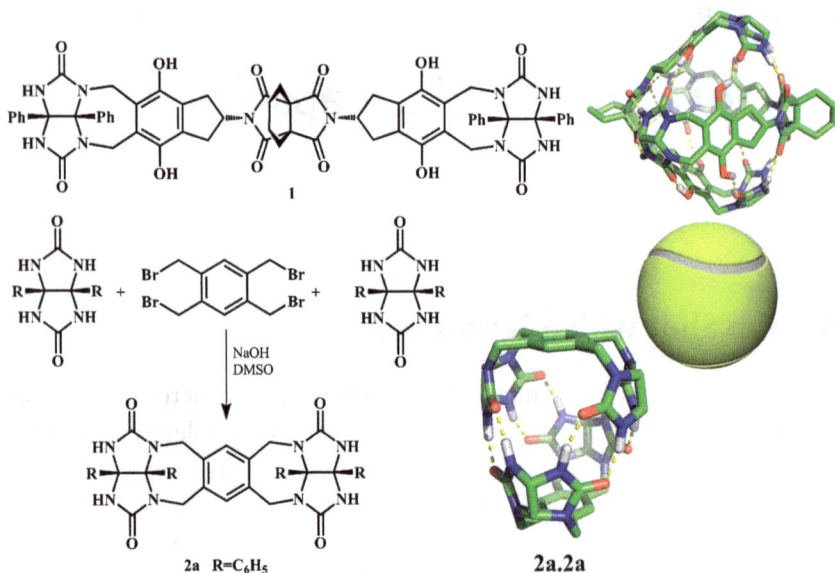

Figure 1.1   *Top*: The original sportsball design and its intended dimerization into a notional tennis ball. *Bottom*: The actual synthesis of the module **2a** and the corresponding hydrogen-bonded capsule with peripheral phenyl groups removed.

observed inside that we were sure of a capsular structure. Our source of methane in the laboratories at MIT showed the presence of considerable ethane and even some propane, but when Neil Branda and Rene Wyler bubbled this gas through a solution of **2a** in CDCl$_3$, encapsulation of the gases occurred (Figure 1.3).[5] The signals for the bound gases at −0.4 and −0.9 represent encapsulated CH$_3$–CH$_3$ and CH$_4$, respectively. The upfield shifts are a result of the anisotropy imparted by the two aromatic spacers of the capsule. The shifts would be even larger but the four peripheral phenyl groups are arranged edge-on to the guest and impart a *downfield* shift. Notice that these signals are sharp and widely separated from those of free ethane and methane, a feature that speaks for a large energy barrier between inside and outside the capsule. In other words, the rate of in–out exchange is slow on the NMR timescale. But exchange is fast on the human timescale, since equilibrium is established upon merely mixing the components. Integration of the spectra gives access to easily calculated

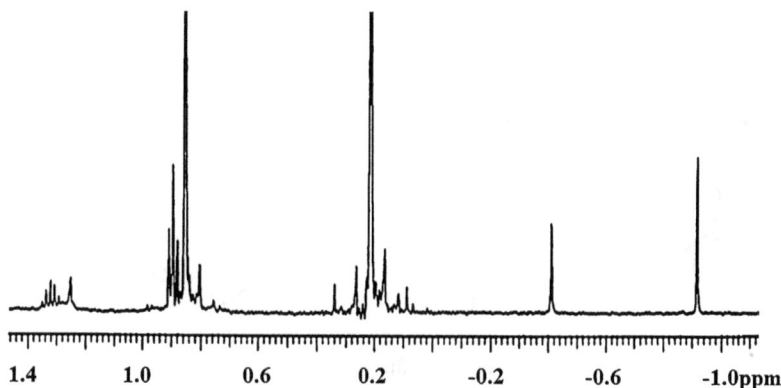

Figure 1.2 NMR spectrum of the tennis ball **2a.2a** in CDCl$_3$ saturated with commercial methane gas. [From *Science* (1994) **263**(5151): 1267–1268. Reprinted with permission from AAAS.]

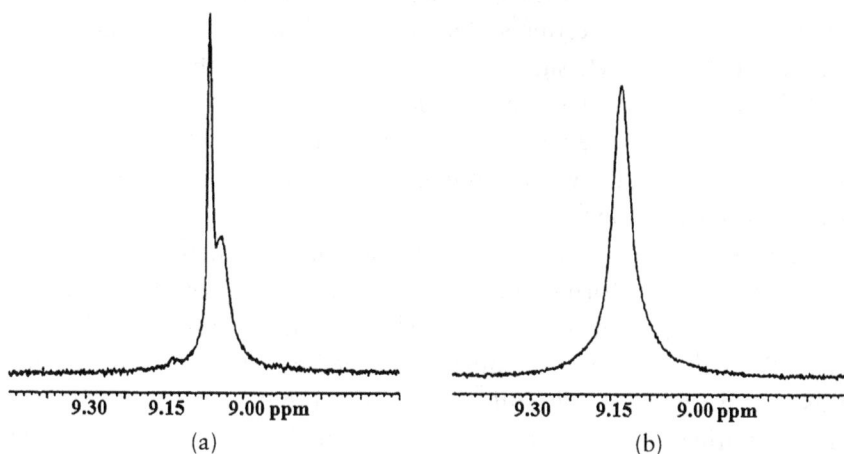

Figure 1.3 Downfield region of the 500 MHz NMR spectrum of **2a.2a** in CDCl$_3$ at −20°C. a) With ambient atmospheric gases; b) after degassing with helium. Reprinted with permission from *JACS* **117**(51): 12733–12745. [Copyright 1995 American Chemical Society.]

equilibrium constants, and changes of temperature give data from which thermodynamic parameters are readily obtained. *We emphasize these features since they are generally observed in hydrogen-bonded encapsulation complexes.*

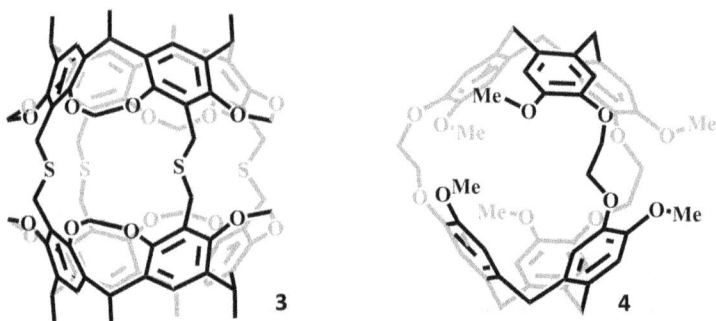

Figure 1.4 Cram's carcerand **3** and Collet's cryptophane **4**, the original covalent container molecules.

The original studies of gas binding in the tennis ball assumed a solvent-filled capsule in $CDCl_3$, but that solvent is much too large to fit inside. Instead, the atmospheric gases dissolved in the solvent are the occupants in the resting state. An example is shown in Figure 1.4, where the N–H NMR signal of the resting state of the tennis ball in $CDCl_3$ shows at least three capsular species present. After flushing with helium, only one species is present.[6] These (NMR-silent) gases are easily displaced by the intended guests, which are presented in higher concentrations.

This should have been a valuable lesson — namely, that encapsulation is best performed in a solvent that cannot compete with the solute for the space inside — but the lesson did not sink in until much later. We will return to this subject, but the importance of solvent size was introduced by Clark Still,[7] who noticed that binding to an open-ended synthetic receptor resulted in very high association constants when a solvent too large to fit was used. Also, and apparently independently, Cram used large solvents that could not be accommodated (incarcerated) by his covalent capsules.[8] The solvent size issue now seems self-evident but was momentous to us then; molecular modeling has vastly improved since then, and solvents of different sizes can quickly be screened for fit. Also, there is the obvious requirement that the solvent be capable of *dissolving* the components, but this has to be done empirically. Finally (for NMR purposes), the desired solvent needs to be available in deuterated form — a requirement that is not always met.

At the time, we were confronted with our first encapsulation complex and needed to place it into some historical perspective. As mentioned earlier, various receptor shapes with concave surfaces, such as clefts,[9] armatures,[10] tweezers,[11] bowls[12] and other vehicles,[13] were under study for recognition. But something special occurs when a guest is completely surrounded by a host. These situations had been encountered with covalent compounds by Cram at UCLA, and Collet in Lyon in the mid-1980s (Figure 1.4). In Cram's case, structures such as 3 became known as carcerands,[14] because in the final step the covalent assembly could capture any nearby, appropriately positioned chemical debris and seal it off — more or less forever, as is characteristic of covalent bonds. In Collet's case there was also a covalent capsule 4, but reversible guest capture was possible.[15] The tennis ball that we had in hand, having had experience with hydrogen bonds and their dynamics of rapid formation and dissipation, was expected to have lifetimes of milliseconds to hours. But determining the lifetimes had to wait for the right expertise to appear in the group. In the meantime, we took a brute force approach to getting things in and out of the tennis ball.

Neil Branda, Robert Grotzfeld and Carlos Valdes prepared more soluble versions of the glycoluril, bearing *p*-dimethyaminophenyl or carboethoxy groups,[16] and incorporated them into the corresponding tennis ball modules 2b,c (Figure 1.5). In deuterated DMF as a solvent, the compounds remained as monomers since the solvent could not be encapsulated. Moreover, DMF competes well for the hydrogen bond donors of the module. But adding gases such as methane, ethylene or even xenon nucleated the assembly of the capsular complex. The

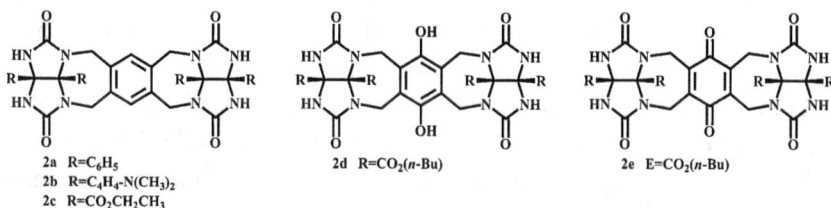

2a R=C_6H_5
2b R=C_4H_4-N(CH$_3$)$_2$
2c R=$CO_2CH_2CH_3$

2d R=CO_2(*n*-Bu)

2e E=CO_2(*n*-Bu)

Figure 1.5 Tennis balls with varied peripheral groups for solubility and proximal groups for recognition of guests.

Table 1.1   Association constants K_a (M^{-1}, 298°K) for the encapsulation of guests in the tennis ball dimers (K_a = [2-guest-2]/[2-2] × [free guest]).

Guest	2c-2c	Host 2d-2d	2e-2e
CH_4	33	70	10
C_2H_6	51	51	13
CH_3F	10	17	<0.3
CF_4	0.7	0.6	<0.2

peripheral basic groups of **2b** were expected to provide sites for protonation and it seemed reasonable that multiple charges on the capsule would force the two halves apart and liberate the guest. The state of the system could be conveniently monitored by both ^{1}H and ^{129}Xe NMR spectroscopy. Control of guest binding and release turned out to be possible in this system: protonation with p-TsOH gave the protonated monomers and free gas, then neutralization with Na_2CO_3 regenerated the capsular assembly **2b.2b**.

The slow exchange of guests allowed the determination of equilibrium constants by simple integration of the NMR signals for free and bound species. These are shown in Table 1.1.

The availability of a variety of peripheral groups on the tennis ball through synthesis paid a great dividend: it eventually led to a crystalline sample, and its X-ray structure was solved by Leticia Toledo. The carboethoxy compound **2c.2c** showed the expected, nearly spherical space inside occupied by a guest.[17] Although disorder prevented the identification of the guest with certainty, it fit the parameters for CH_3OH, one of the solvents of crystallization. A stereoview of the capsule with deleted ester groups is shown in Figure 1.6(A), while the disorder of the esters can be seen in Figure 1.6(B). This was the first direct observation of a solvent molecule inside, albeit in the solid state. The dimeric structure showed signs of binding CH_2Cl_2 but was reluctant to take up the larger $CHCl_3$. The use of $^{13}CH_2Cl_2$ permitted the direct observation of this guest inside and in solution for the first time.

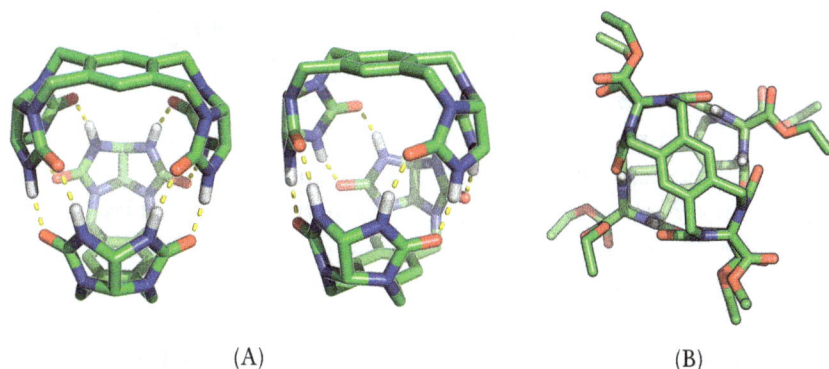

Figure 1.6   (A): Stereoview of the solid state structure of the tennis ball **2c.2c** with the carbethoxy groups removed. (B): A view of the dimer showing the disorder of the ester groups.

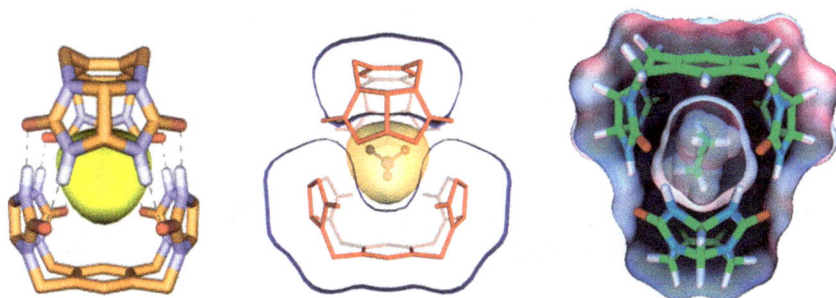

Figure 1.7   Views of the tennis ball. *Left*: Hydrogen bonds. *Center*: solvent accessible surfaces of capsule and CH_4 guest. *Right*: Modeled complex with C_2H_6.

What are the relative sizes of the guests and the cavity of the tennis ball? It is easy to be misled by software in answering this question. What you see depends on the software you use, as illustrated by Figure 1.7. Simple modeling packages are accurate in finding the eight hydrogen bonds but rolling a standard H_2O-size probe around to find the solvent-accessible surface of the tennis ball and methane leads to the central figure. The surfaces overlap and one could conclude that methane is too big to fit inside. More sophisticated software gives the rendering on the right, which shows that even ethane leaves plenty of empty space inside. Other professional computations of the capacity followed.[18]

The self-complementary hydrogen bond patterns of glycoluril demanded incorporation into related architectures with different spacers. A shorter one, ethylene (as in **5**, Figure 1.8 (top)), was clearly a good candidate for assembly into a dimeric capsule, as its O-to-O dimensions were close to those of the tennis balls. The longer spacers, such as naphthalene **6** and bridged anthracene **7**, showed less promise, but these were all eventually synthesized by Carlos Valdes, Uri Spitz and Stefan Kubik. The ethylene-spaced module **5** ($R = CO_2Et$) formed

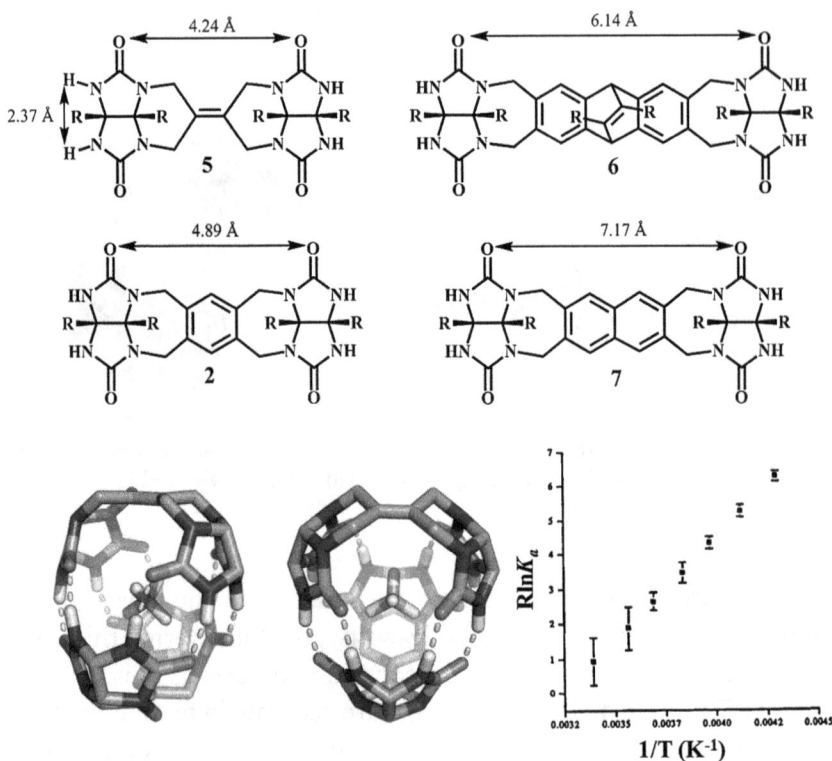

Figure 1.8 *Top*: Glycolurils separated by different spacers also dimerize. *Bottom*: X-ray stereoview of dimer **5.5** without esters and the van 't Hoff plot for its binding of CH_4 in $CDCl_3$. [Reprinted with permission from *Chemistry — A European Journal* 2(8):989–991. Copyright 2006 Wiley-VCH Verlag GmbH & Co. KGaA, Weinheim.]

a capsule **5.5** (Figure 1.8) readily and a crystalline sample gave the expected structure shown.[19] The new capsule showed a desirable property: it selectively bound methane, albeit weakly, but rejected ethane and larger gases. The K_a for CH_4 in $CDCl_3$ under ambient conditions was only 3.8 M^{-1} and the van 't Hoff plot gave the parameters $\Delta G = -0.7$ kcal/mol, $\Delta H = 5.7$ kcal/mol and $\Delta S = 18$ cal/mol. The energetics resemble classic solvophobicity: entropy-driven assembly with massive release of solvent. With its volume, CH_4 filled about 50% of the space, whereas ethane filled more than 80%. Gases are expected to be even more space-demanding, and have much lower occupancy factors, as we will relate later. The calculated packing coefficient of greater than 80% for CH_3–CH_3 would be prohibitively high for a gas. Typically, packing coefficients, even in the solid state, are around 74% (the figure for closely packed spheres).

The larger **6** and **7** did assemble weakly as homodimers but they did better as components of heterodimeric structures. While they assembled, their larger holes were not able to retain guests, at least with the slow exchange rate that would allow direct observation inside. Like Reinhoudt's "Holand,"[20] typical guests the size of solvent molecules drifted in and out rapidly. They spent little time inside and their spectra did not change enough, i.e. the environment inside and outside were averaged and time-weighted. This is a general feature of open-ended receptors, the most famous being the cyclodextrins; it becomes very hard to interpret the details of complex structure when rapid exchange on the NMR timescale is observed.

However, the NMR characteristics of the modules **5**, **6** and **7** changed as a function of the proffered guests and the binding of solvents could be deduced. For example, the NMR signals of **7** in the (too large) $CDCl_2CDCl_2$ sharpened on the addition of the somewhat better guest $CDCl_3$, and the downfield resonance of the N–H signal at 8.25 ppm heralded the formation of the capsule **7.7**. The likely heterodimeric possibilities with these spacers are shown in Figure 1.9. All of these were observed when the appropriate solvents were present to fill the cavities and the respective disproportionation constants K are given in Table 1.2.

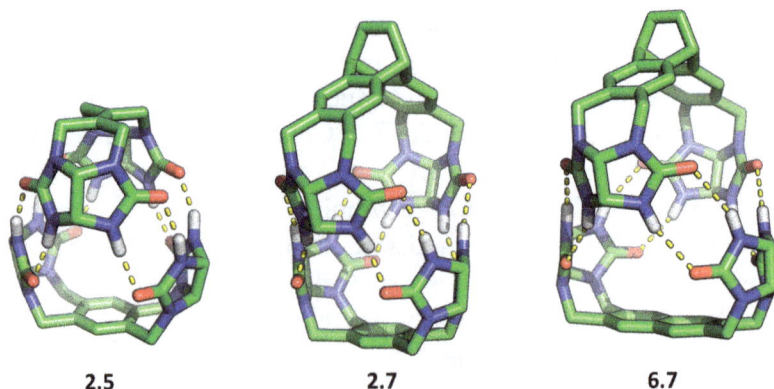

2.5   2.7   6.7

Figure 1.9 Hybrid capsules formed by heterodimerization of modules bearing gly-colurils on different spacers.

Table 1.2 Equilibrium (Disproportionation) Constants, K,[a] for the Formation of Hybrids (Heterodimers) in Binary Mixtures using Solvents of Increasing Size, and Thermodynamic Values, **ΔH** and **ΔS**, for the Equilibrium in CDCl$_3$.

Hybrid dimer A • B	$K_{298} = \frac{[A*B]}{[A*A][B*B]}$				ΔH (kcal/mol) CDCl$_3$	ΔS(eu) CDCl$_3$
	CD$_2$Cl$_2$	CDCl$_3$	CDBr$_3$	CDCl$_2$CDCl$_2$		
2 • 5	0.08	0.9	0.7	0.9	2	8
2 • 7	No 2 • 7	11.6	42	Trace of 2 • 7	−3	−5
6 • 7	0.2	5.7	62	0.6	5	21

[a]Determined by ^{1}H NMR by integration of the N–H signals of the different species (errors are estimated as ±10%).

The purely statistical value for disproportionation is 4, and that is nearly reached with CDCl$_3$ in the **6.7** combination. This result indicates a rather indifferent fit of this guest in the **6.6** or **7.7** combination. The larger *K* (11.6) for this solvent in **2.7** heralds its better fit in that heterodimer. The sizable numbers for CDBr$_3$ in the heterodimers **2.6** and 2.7 indicate a good fit, while the CDCl$_2$CDCl$_2$ appears too large for any combination. The guest CH$_4$ showed signals at −0.42, −1.5 and −1.0 in the capsules **5.5**, **2.2** and **2.5**, respectively. The larger lesson here is the promiscuity of the hydrogen bond pattern; self-assembly usually involves corrections that discriminate between self and non-self, but these systems appear to have no such inhibitions. This was our first encounter with the notion of self-sorting, and how it does not work with our hydrogen-bonded systems. Again, this lesson did not sink in until later. Had we explored irregularly shaped guests at the time, we may have had a shortcut to discoveries that appeared only a decade later.

Before we leave the tennis ball, a description of its energetics and dynamics is in order. This had to wait some three years until the appropriate molecules could be accessed and the relevant expertise appeared in the group. The desymmetrized module **2f**, bearing different glycoluril peripheries (Figure 1.10), assembles into a chiral tennis ball. By this we do not mean a chiral space but only an object without a plane of symmetry or center of inversion. The dimeric **2f.2f** assembles around a gas as a pair of enantiomers which can interconvert or racemize only by dissociation to the modules and then their recombination. This process exchanges chemical environments of the nonequivalent N–H signals at each glycoluril in the NMR spectrum. The NMR can also measure the exchange between bound and free guest, provided, of course, that the rates occur on the NMR timescale. Fortunately, they did so.

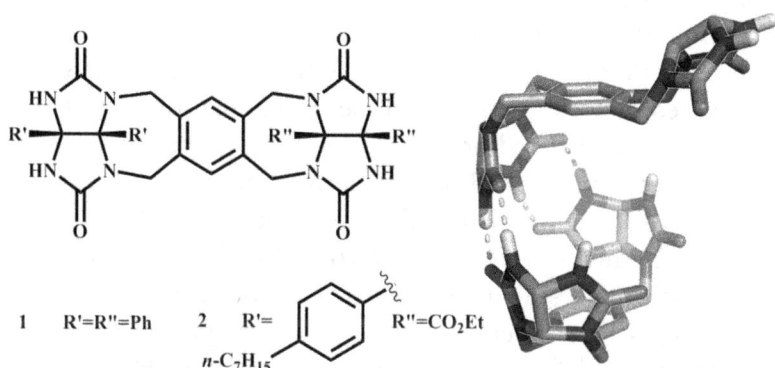

Figure 1.10 *Left*: A module with 2 different glycolurils leads to a chiral (but racemic) tennis ball dimer. A partially opened tennis ball allows guest exchange without dissociation of the dimer (peripheral groups have been deleted).

Table 1.3 Kinetic Data[a] for the Dissociation/Reassociation Rates [$k_{diss} = 2k_{rac}$] of 2f • 2f and Exchange Rates [k_{ex} and $k_{ex'}^-$] for Guests Encapsulated by 2f • 2f.

Guest	k/s^1	$\Delta G_{295}^{\ddagger}/(\text{kcal mol}^{-1})$
empty[b]		
k_{diss}	0.62	17.5 ± 0.1
methane		
k_{diss}	0.14	18.5 ± 0.3
k_{ex}	1.14	17.2 ± 0.1
$k_{ex'}$	1.0	17.3 ± 0.1
ethane		
k_{diss}	0.10	18.6 ± 0.1
k_{ex}	0.56	17.6 ± 0.1
$k_{ex'}$	0.46	17.7 ± 0.1

[a]Reported values are the average of three sets of measurements. [b]Empty or occupied by dissolved atmospheric gases.

Thomas Szabo and Göran Hilmersson prepared the appropriate unsymmetrical tennis ball. They did the EXSY experiments in CDCl$_3$ and determined the rates of capsule dissociation k_{diss} (which are defined as 2× the racemization rate, k_{rac}); the rates of guest exchange, k_{ex}, for methane and ethane, and the difference, $k_{ex} = k_{ex} - k_{diss}$ (Table 1.3).[21] Since guest exchange is faster than dissociation, *there is a process that allows in–out motion of the gas without complete dissociation of the tennis ball.*

An estimate of the dimerization constant K_a can be made from the measured dissociation rate and the recombination rate (assumed to be diffusion-controlled or 2×10^9 M/s under these conditions).[22] For the "empty" tennis ball (occupied by atmospheric gases) the K_a is then 3×10^9 M^{-1} (or dissociation constant $K_d = 0.3$ nM).

The energy involved is $\Delta G° = 12.6$ kcal/mol, a value that agrees well with the rough estimate of about a factor of 10 for each hydrogen bond[23] in CHCl$_3$, although secondary effects[24] may also be in play in the case in hand. The encapsulation of CH$_4$ or C$_2$H$_6$ adds about 1 kcal/mol to this value, presumably through CH–π interactions.[25]

An intermediate for guest exchange without capsule dissociation is proposed in Figure 1.10. Inversion of a seven-membered ring and breaking four hydrogen bonds creates an opening large enough to accommodate the passage of small gases. The four hydrogen bonds that remain holding the system together should permit a lifetime in the microsecond range — plenty of time for exchange of guests to take place. These could occur as either S_N1 (guest dissociation) or S_N2 (guest substitution) processes. We will return to the guest exchange details with the softball.

We next considered triphenylene as a spacer, but this cleary required adjustments of architecture. In expanding the width, compensating changes in another dimension had to be made. Modeling indicated a beautiful fit of a structure such as **8.8** with three glycolurils on the triphenylene spacer (Figure 1.11). The molecule required a great deal of synthetic work, which was eventually completed by Robert Grotzfeld and Neil Branda.[26] The resulting structure — say, on the scale of the tennis ball — would resemble something the size of a hamburger: a flattened sphere that had a capacity for disk-shaped guests.

Benzene was indeed a superb guest, but the most informative guest was cyclohexane. That guest had the appropriate dimensions to make attractive CH–π interactions with the aromatic panels above and below. This snug fit caused a decrease in the rate of the ring flip of encapsulated cyclohexane. The rates were measured by Brendan O'Leary and Robert Grotzfeld with cyclohexane-d_{11}. The axial and equatorial hydrogen resonances show huge chemical shift differences inside this capsule (Figure 1.11) and the accuracy of the rate determination was insured by running the same experiment *outside of the capsule at the same time, in the same NMR tube.*[27] While the effect is small, only about 0.3 kcal/mol, the CH–π contacts between guest and host must be broken as cyclohexane goes to twist-boat, and then to the transition state for the ring flip. This was our first observation of a peculiar behavior in a small space — molecules within molecules[28] — versus that behavior free in bulk solution.

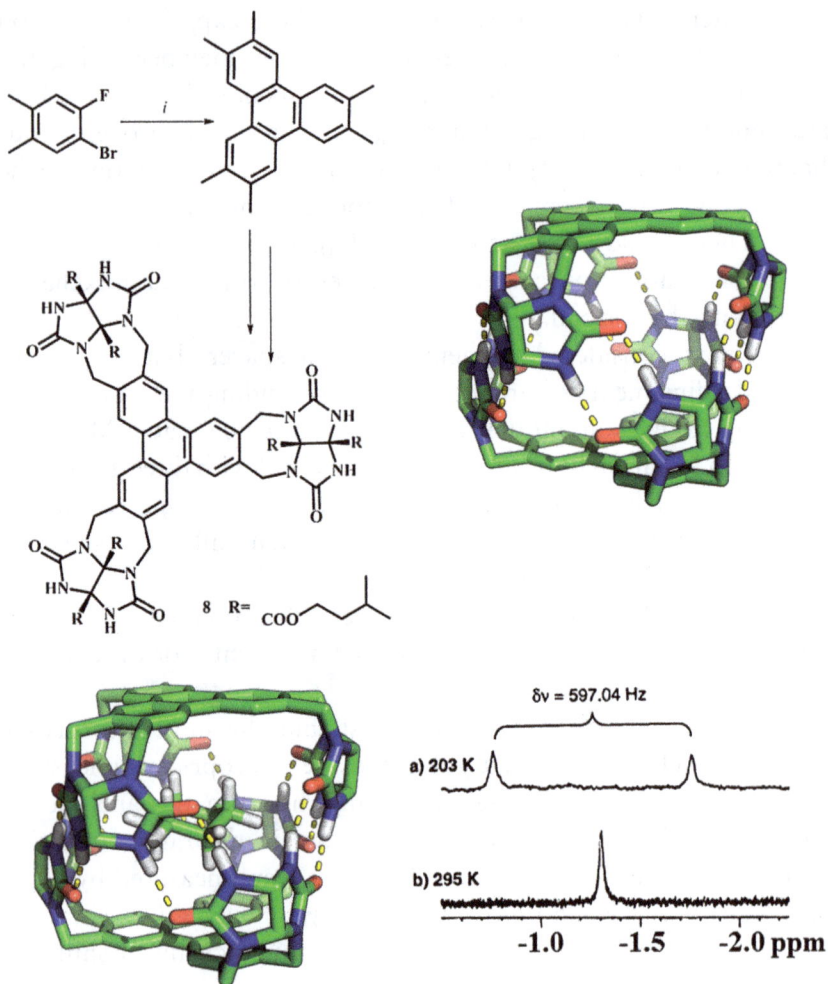

Figure 1.11 *Top*: Synthesis of the jelly doughnut **8** and its dimeric capsule **8.8**. *Bottom*: Encapsulated cyclohexane and the partial ^{1}H NMR spectra of cyclohexane-d_{11} in the capsule. [Reprinted with permission from *J. Am. Chem. Soc.* **119**: 11701–11702. Copyright 1997 American Chemical Society.]

At about the same time, we returned to the original target — the softball. Rob Meissner had taken over the synthesis of **1**; it is outlined in Figure 1.12. On its completion, it did not behave as expected. Rather, the NMR spectra could only be interpreted as a molecule collapsed onto itself with intramolecular hydrogen bonding.

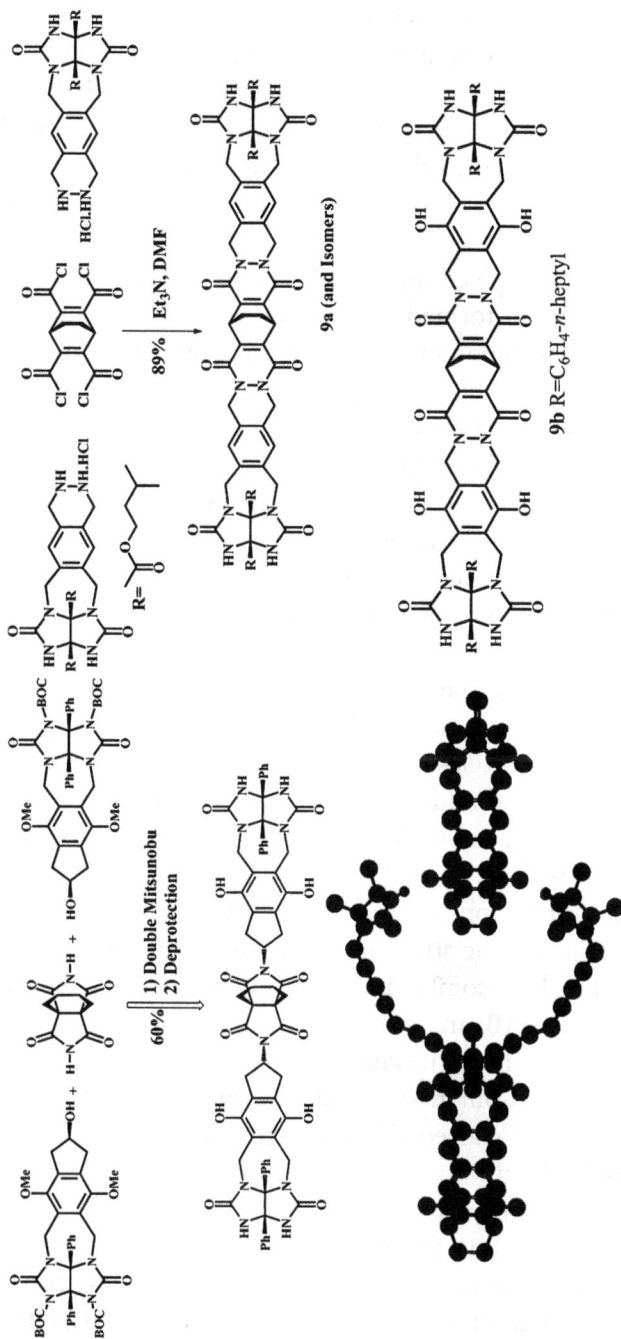

Figure 1.12 *Top*: Syntheses of **1** and the more rigid **9a** softball. *Bottom*: Proposed chain-like assembly of **9a** in CDCl₃ that produces a gel phase but dimerizes into the softball in benzene. The more stable and soluble **9b**.

The freely rotating sigma bonds in the structure were deemed to be the culprits.[29] Corrections were duly made by Meissner, who converted the basic skeleton into the more rigid system **9a** (Javier de Mendoza also suggested the same improvements). At any rate, the new synthesis involved 19 steps and led to the compound with 13 fused rings and one bridged ring system to impart the rigid fold needed in the center of the structure.[30]

The molecule dissolved in $CDCl_3$ but after a few moments an intractable gel was formed, indicating some type of aggregate. A proposed structure, involving a head-to-tail oligomer, is shown, but there is no evidence for this or any other assembly. Rather than succumb to cruel disappointment, Meissner tried aromatic solvents — evidently a better fit — and found success. The compound **9a** presented sharp NMR signals with well-resolved peaks in C_6D_6 and also C_6D_5F, but these spectra did not reveal whether the compound remained a monomer, dimer or some higher aggregate.

However, in a mixture of C_6D_6 and also C_6D_5F, a third set of signals appeared which indicated that one of each solvent species was encapsulated. Given the size requirements for two such solvents, a dimeric structure was indicated. With an ill-fitting solvent such as *p*-xylene, a broad featureless NMR spectrum was obtained, but as soon as a good guest was added, such as tetramethyl adamantane, the sharp spectra of a single complex emerged. We referred to this process as "autoencapsulation." For reasons of greater solubility **9b** was synthesized, and for reasons of greater stability OH groups were picketed along the edges of the structure to provide additional hydrogen-bonding sites. The best combination for solubility and stability was found in the structure **10**, and most of the subsequent studies with the softball were done with this module.

Although this was our first encounter with two coencapsulated guests, right away we started thinking about bimolecular reactions inside the capsule. Naturally, a cycloaddition reaction was the most obvious choice, and this was eventually pursued,[31] and will be discussed in Chapter 8. But the inclusion of two solvent molecules made for some very unusual thermodynamic behavior we encountered early in the study of guest inclusion in the softball. Encapsulation of,

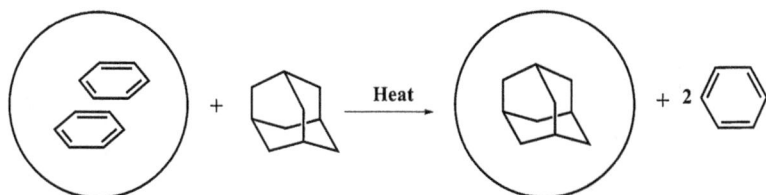

Figure 1.13   Entropy-driven encapsulation of adamantane as 2 resident benzenes are displaced.

say, adamantane displaces two benzenes and as the system is heated more adamantane goes in. Jongmin Kang found that the encapsulation of adamantane is entropy-driven,[32] as the equation in Figure 1.13 shows: there are two particles on the left side of the equation and three on the right.

The exchange of adamantane with 2,2-paracyclophane in the softball took place at convenient rates at NMR concentrations, and permitted extensive and conventional kinetic studies to be performed on this system. The cyclophane is a much better guest, so the displacement is essentially irreversible, and the half-life is about 40 min at millimolar concentrations. We dwell on the mechanism at length because it applies to almost all of the capsules we have looked at. In considering how the reaction occurs, and its rate-determining step, one might propose pulling apart the two halves of the softball — i.e. breaking all of the hydrogen bonds, and whatever bonds or intermolecular attractions there are between host and guest — then rinsing out or exchanging the guest in whatever half it is in, and reassembling the system. This, of course, would be energetically the most costly path, and therefore the least likely. Moreover, as we will see from later experiments, the exchange of guests is always faster than the exchange of capsule modules in these assemblies. Instead, consider opening opposite flaps of the softball as in (Figure 1.14(A)). This requires mere inversion of two six-membered rings, but it does entail the breaking of 12 hydrogen bonds. Still, four hydrogen bonds hold this system together, and an appealing opening can be envisioned for an inline S_N2-like displacement with such an intermediate.

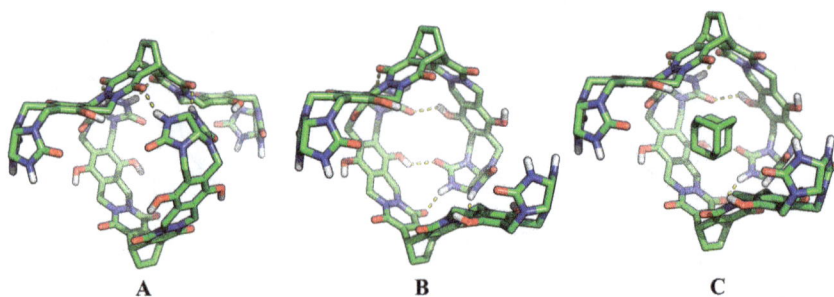

Figure 1.14   (A): Opening of opposite "flaps" exposes the interior of the softball to in-line entry and exit. (B): Opening of adjacent flaps conserves more hydrogen bonds. (C): Adamantane as the guest in an opened softball.

Alternately, consider the related but energetically less costly path that opens two adjacent flaps of the softball, as in Figure 1.14(B). This intermediate provides enough of an opening for a displacement of, say, an adamantane resident guest to occur with six hydrogen bonds holding the assembly together. Such an intermediate as in Figure 1.14(C) would have a lifetime of milliseconds in a noncompeting solvent like toluene — long enough for the substitution to take place.

The experiments were undertaken and interpreted by Javier Santamaria, Tomas Martîn, Göran Hilmersson and Stephen Craig, who were able to determine the various rate and equilibrium constants involved.[33] The upshot is shown in the figure, and the mechanism, S_N1- or S_N2-like, depends on the conditions. At high concentrations of the incoming guest (paracyclophane), the rate-determining step is the opening of the flaps, and each time the flaps open, the displacement occurs. The rate is independent of the incoming guest's (high) concentrations. At low concentrations of the incoming guest, the flaps open and close many times before the displacement occurs, in more of an S_N2-type process. The rate increases with the incoming guest's concentration. We will see later how this allows a chiral space to be assembled with a chiral guest and then maintained long after the guest is gone. A computational analysis by Houk refers to exchange through a series of "gating" mechanisms and proposes alternative but complementary sequences.[34]

The tennis ball and other spherical capsules involve dimerization of self-complementary structures; the curvature and hydrogen-bonding patterns provide instructions for assembly. A more complex system with four identical components was desirable and we decided to cut each piece of the tennis ball in half, as represented below; in other words, desymmetrize the components (Figure 1.15). In terms of its chemical implications, this was accomplished by keeping the tennis ball's curvature but engineering the hydrogen bond patterns to force assembly of components in a head-to-tail manner.

The first structure **10** (Figure 1.16), synthesized by Carlos Valdes and Leticia Toledo, had the appropriate patterns but lacked an enforced curvature. The urea function was dynamic: it could move above or below the aromatic spacer's plane, and the X-ray structure showed a ribbon-like structure instead of the desired closed shell.[35]

Tomas Martin and Ulrike Obst[36] corrected this flexibility by removing the two methylenes and replacing the urea with a superior hydrogen bond donor — a sulfamide **11**. The second-generation structure maintained the glycoluril's superior hydrogen bond acceptors with the aryls on the molecule's convex front. The figure shows

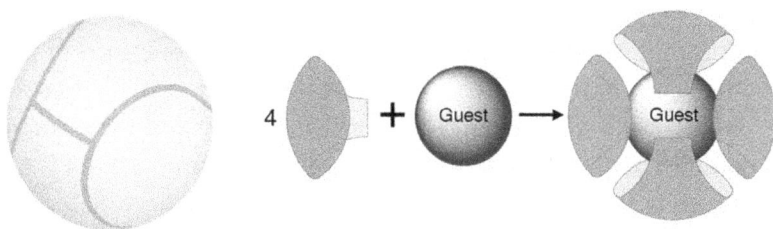

Figure 1.15 *Left*: Bisection of the notional tennis ball. *Right*: Head-to-tail assembly of 4 modules around a guest.

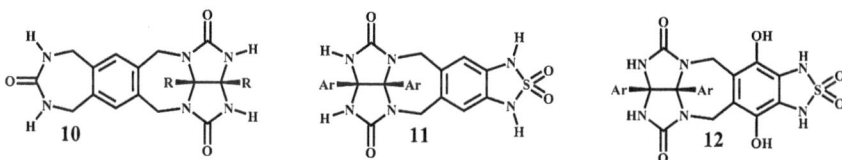

Figure 1.16 Desymmetrized modules bearing enhanced hydrogen bond donors and acceptors.

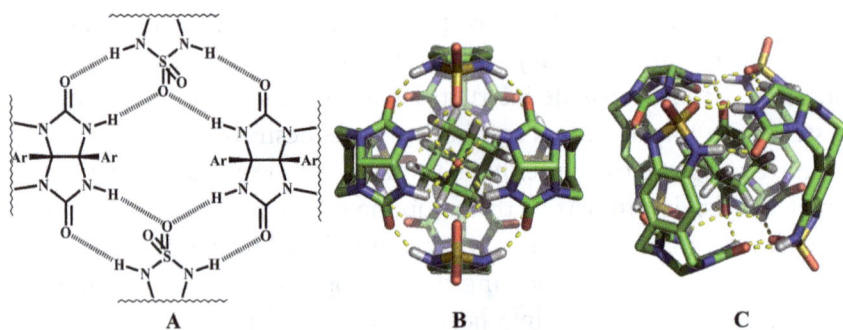

Figure 1.17 (A): Hydrogen bonding between sulfamides and ureas. (B): The tetrameric capsule with adamantanedione as guest. (C): A cutaway view of the host–guest assembly.

an outer circle of the best hydrogen bonds and a default inner circle of glycoluril N–H weak donors matched with sulfamide's weak S–O acceptors.

This module assembled as an achiral tetramer with suitable guests such as adamantane ketones. The carbonyls of the guest can make bifurcated hydrogen bonds with the seams of the host. A third-generation derivative **12**, synthesized by Fraser Hof, Peter Iovine and Colin Nuckolls, had additional attributes: the hydroxy groups of each monomer donate one intermolecular and one intramolecular hydrogen bond to the assemblies. A derivative provided crystals suitable for x-ray analysis and Darren Johnson solved the structure. The complex with adamantanedione, a perfect symmetry match for the capsule, featured the expected hydrogen-bonded network.[37] Each carbonyl oxygen atom of the diketone guest accepts hydrogen bonds from four identical N–H donors of the host, as shown. The module was further desymmetrized to give a chiral structure, as described in Chapter 6.

Sieves

A larger version combined the glycoluril module, for molecular curvature and hydrogen-bonding sites, with a preorganized platform: the hexa-substituted benzenes.[38] The geared arrangement of substituents around the benzene ring is much admired as a means of presenting all three modules on one face of the spacer, as shown in the structures **13** (Figure 1.18).[39]

The dimeric capsules were detected in the gas phase by mass spectrometry (MALDI) as homodimers. In solution, NMR was used to detect the formation of heterodimeric assemblies. Mixing of the two modules — which differed only by the "direction" of the amide bonds — gave a practically statistical distribution of dimeric species, and the heterodimer is modeled in Figure 1.18. Nonetheless, the amide directionality had serious consequences for guest encapsulation. The heterodimer structure indicates that the amide carbonyls that are directed into the cavity occupy much more space than the isomeric arrangement shown, where the amide N–Hs are directed inward. Consequently, the homodimeric capsules have different cavity sizes and shapes and different guest affinities. The sizable holes

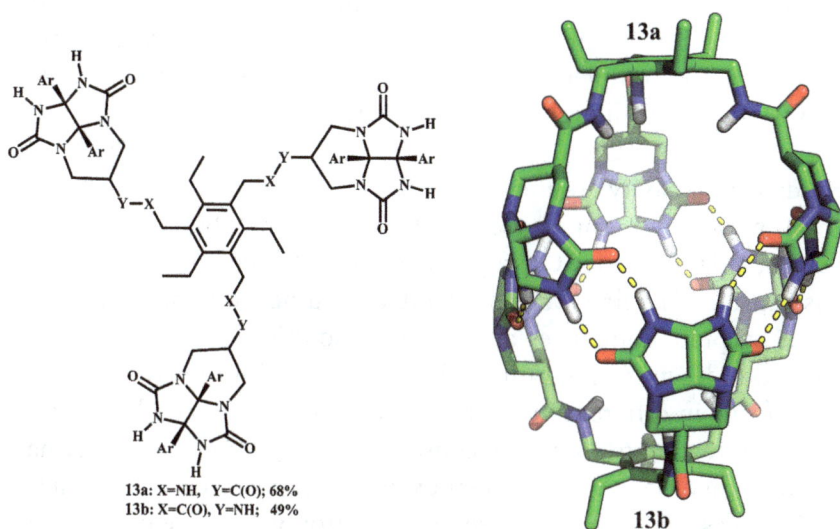

13a: X=NH, Y=C(O); 68%
13b: X=C(O), Y=NH; 49%

Figure 1.18   *Left*: The modules of the molecular "sieves." The geared arrangements of the groups on the benzene present all 3 glycoluril modules on the same face of the structure. *Right*: A model of the homodimer **13a.13a**.

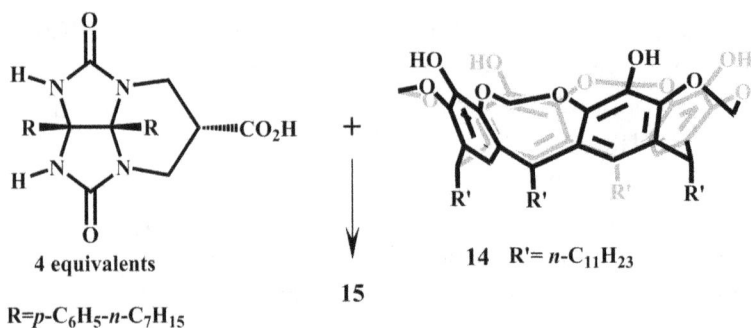

Figure 1.19   Synthesis of a molecular sieve using Cram's resorcinarene platform.

in the structure allow rapid exchange of small (solvent) molecules, but larger guests (paracyclophane, ferrocene) exchange slowly on the NMR timescale.

A second sieve-like capsule was fashioned from the same glycoluril module but placed on a shallow cavitand platform **14** (Figure 1.19).[40] As we will relate below, this platform is the most famous of shallow cavitands and is widely applied in capsule formation; it was also the basic element in Cram's carcerand synthesis.[14] In the present application, the self-complementary shape of the C_{4v}-symmetric monomer **15** would permit dimerization into a D_{4d}-symmetric capsule **15.15**, shown in Figure 1.20. The capsule was held together by 16 hydrogen bonds with a cavity volume of $\sim$950 Å^3.

A number of unprecedented features were observed by Arne Lützen, Adam Renslo, Cristoph Schalley and Brendan O'Leary with this capsule. The first was its size: it was sufficiently large to encapsulate cryptate complexes as shown. In solution, the NMR spectra showed a desymmetrization of the complex, with the two halves of the capsule in different magnetic environments. The combination of the cryptate and its anion can cause this: one occupies the upper half and the other the lower, and their exchange of positions is slow on the NMR timescale. This restricted guest motion within the capsule was unexpected, and further experiments were conducted to confirm the interpretation. Complexation of the ion pair K^+SCN^- was examined by using isotopically labeled salt $K^+S^{13}CN$. In the presence of the capsule, an intense signal for the $S^{13}CN$ anion arose at 132.5 ppm,

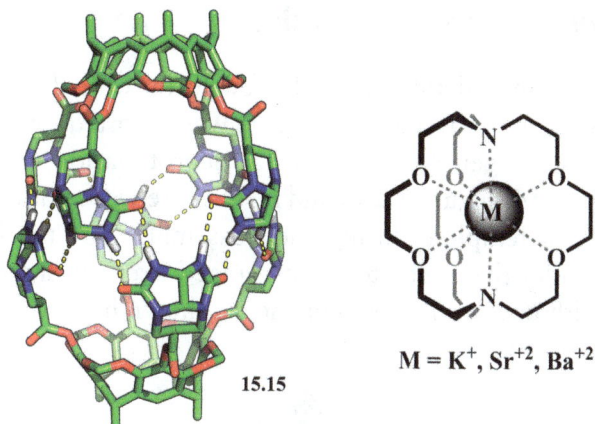

15.15

M = K⁺, Sr⁺², Ba⁺²

Figure 1.20 *Left*: A molecular "sieve" that allows small solvents to move in and out rapidly. *Right*: Cryptate guests move in and out of the sieve slowly and act as labels for characterization by mass spectrometry.

indicating that the ion pair is solubilized by encapsulation; the salt alone was not able to give a signal, owing to its insolubility. In contrast, the much larger salt K⁺ B(*p*-ClPh)₄ gave no sign of encapsulation despite its greater solubility.

The second new departure was made in gas phase studies with electrospray ionization mass spectrometry (ESI-MS). The team used the cryptates not only as the guests, but also as the ion labels necessary for ESI-MS characterization of the capsule (6890 amu). The mass range of the instrument was limited (*m/z* < 4000 amu) but the use of cryptate salts with doubly charged cations Ba²⁺ and Sr²⁺ permitted the observation of the complexes. The introduction of ammonium ions as guests *and* ion labels revolutionized our ability to observe other capsules in the gas phase.

The third new aspect is what may be called a "second-sphere" supramolecular system: a complex within a complex. This double inclusion was rare in the host–guest literature but examples existed in the solid state involving complexes of crown ethers[41] and cryptates[42] by cyclodextrins. The phenomena in the solution and gas phases were novel and invoked Matryoshka dolls. Even larger capsules will lead to more frequent and more elaborate structures in the future.

Other Resorcinarene Modules

The earliest form and still one of the few forms of hydrogen-bonded capsules involving charged components was reported by Sherman.[43] The capsules are generated by partial deprotonation of the resorcinarene tetrol **14** (Figure 1.21), and reversible encapsulation of guests was used as a template during their assembly. For the hydrogen-bonded capsules the template effects ranged over six orders of magnitude,[44] with guests such as pyrazine, as shown in **16** (Figure 1.21).

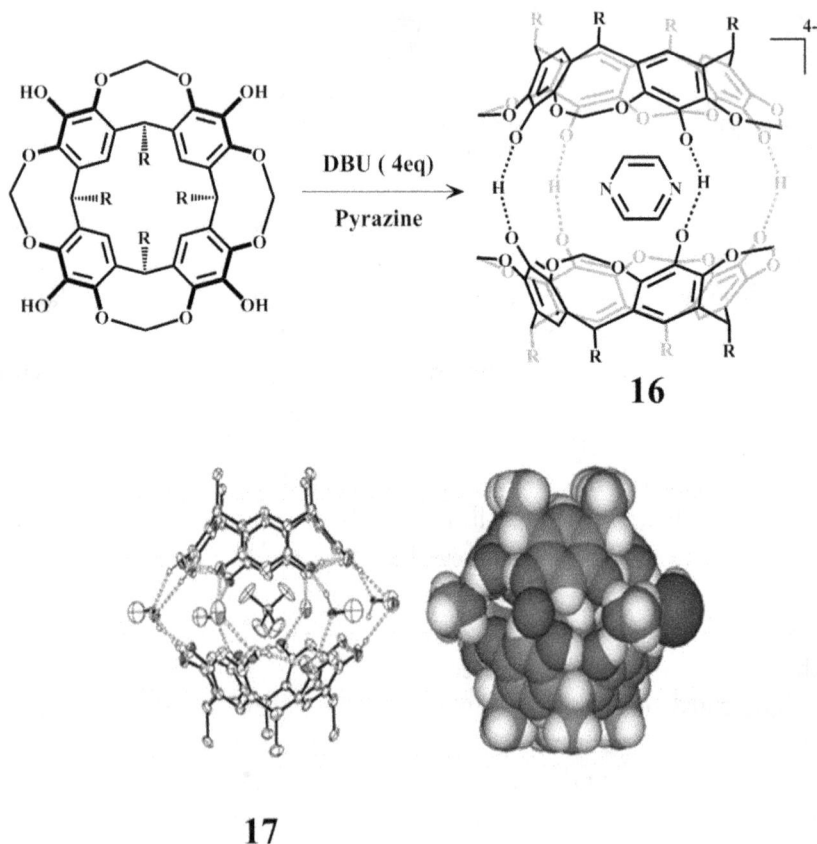

16

17

Figure 1.21 *Top*: Partial deprotonation of the tetrol in the presence of guests templates the formation of capsules. *Bottom*: Deprotonation is not required for encapsulation as shown for $Me_4N^+@1_2 \cdot Br^- \cdot 4MeOH \cdot 3H_2O$ in **17**.

These effects are all the more remarkable for their stability in DMSO-d_6 as solvent, since DMSO is a strong competitor as a hydrogen bond acceptor.

Hydroxylic solvents can participate as bridges between the resorcinarene hemispheres, and the need for deprotonation for capsule formation can be bypassed entirely.[45]

The capacity of the basic shallow cavitand was further expanded by Kobayashi.[46] Divergent functions attached to the resorcinarene carboxylic acid **18** (Figure 1.22) were matched with spacer elements (2-amino-pyrimidines) providing hydrogen bond and shape complementarity. Two molecular hemispheres and four spacers came together in a six-component capsular assembly. The capsule accommodated sizable aromatic guest molecules, such as 2,6-dimethoxynaphthalene.

Extensions of the architecture by the insertion of aryl groups as in **19** (Figure 1.23) made for large-capacity capsules, and a combinatorial diversity of assemblies were available through a mix-and-match strategy.[47] Complexes were characterized by ^{1}H NMR titration and x-ray crystallographic analysis, and revealed aspects of guest orientation control. The positioning of functional groups lining the space of the heterodimeric assemblies augurs well for the application of these capsules as reaction chambers.[48]

Additional use of the resorcinarene cavitand as a platform was made by Paek.[49] He described hydrazides as self-complementary components for hydrogen-bonded capsules and expanded these to derivatives of hydantoins **20** (Figure 1.24).

The appropriate elaboration of the hydantoins with, say, α,α disubstitution could give a capsule (if it assembled) with inwardly directed functional groups and possibilities for chiral spaces inside. Further extension of the resorcinarenes with acetylenic pyridine spacers gave modules **21** described by Aakeroy[50] (Figure 1.25). Their hydrogen-bonded dimers had capacity for three ethanol molecules, two methanol molecules, and a toluene molecule inside.

An architecturally distinct self-assembling module was also synthesized by de Mendoza, who used the curvature of cyclotriveratrylene (CTV) as the platform.[51] The CTV was introduced by Collet in his synthesis of the cryptophanes — one of the original covalent

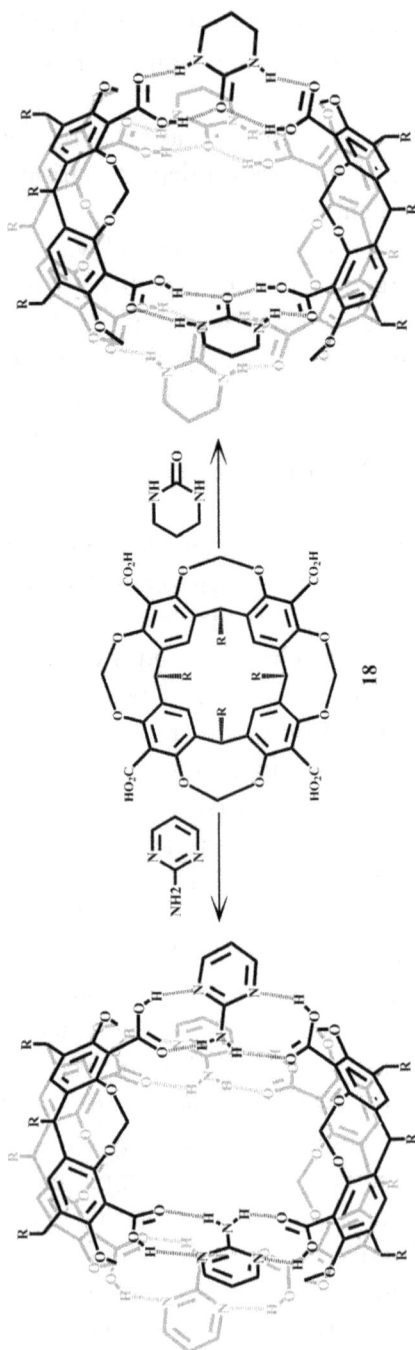

Figure 1.22   A tetracarboxylic cavitand forms capsules with spacers that offer complementary hydrogen-bonding sites.

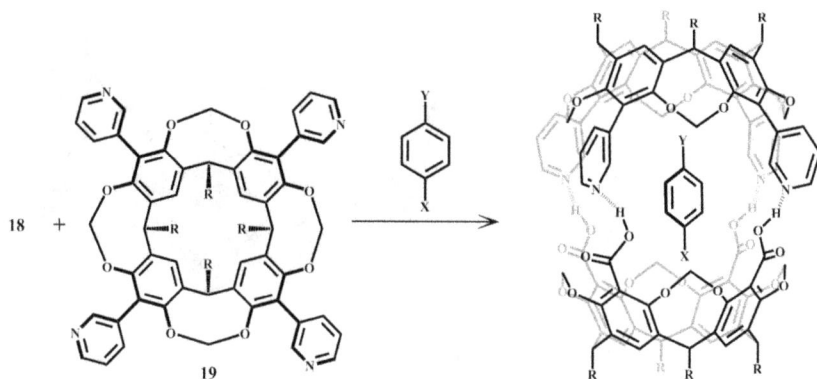

Figure 1.23   *Left*: A resorcinarene cavitand extended with pyridines. *Right*: A heterodimeric capsule held together with hydrogen bonds featuring strong acid/base character.

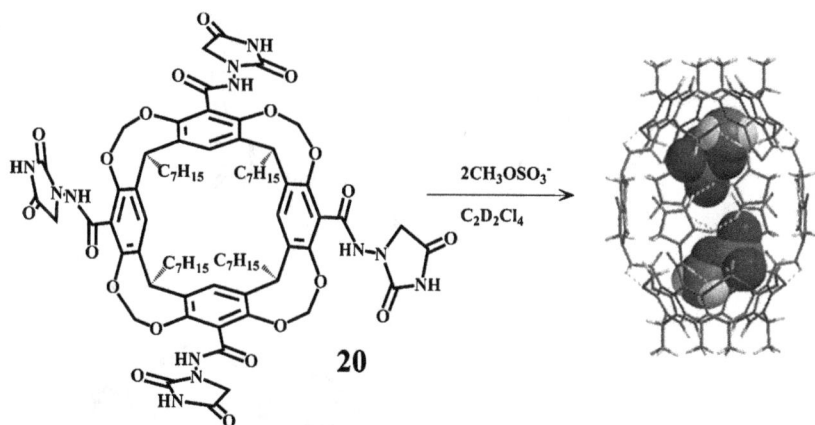

Figure 1.24   Self-complementary hydantoins on the resorcinarene platform dimerize and encapsulate 2 mesylate anions. [Reprinted with permission from *Org. Lett.* 10: 4867–4870. Copyright 2014 American Chemical Society.]

capsules.[15] The positioning of three hydrogen-bonding units on the CTV led to **22** and then dimerization into a pseudospherical capsule (Figure 1.26). The volume of nearly 800 Å^3 suits some of the higher fullerenes. The capsule was applied in solid–liquid extractions to purify C$_{70}$ from crude soot and to obtain mixtures enriched in

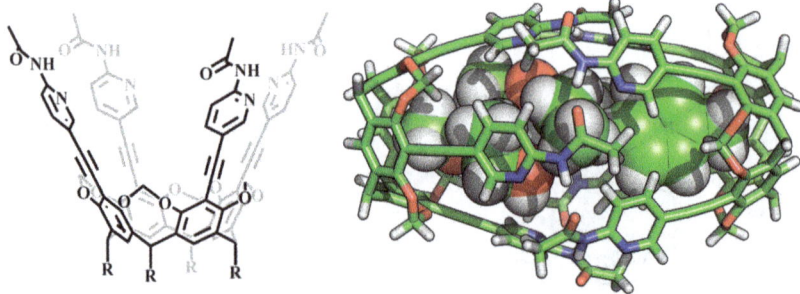

Figure 1.25   Self-assembly of an expanded capsule with 5 guests inside.

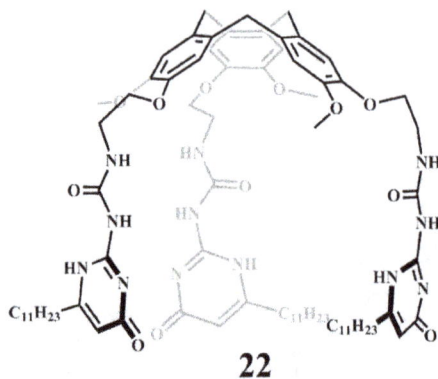

22

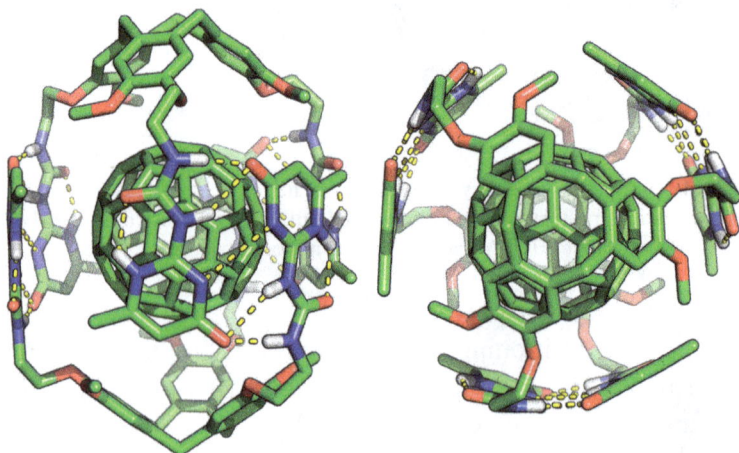

Figure 1.26   *Top*: A cavitand derived from cyclotriveratrylene. *Bottom*: Two views of the dimeric capsule with C$_{70}$ as guest.

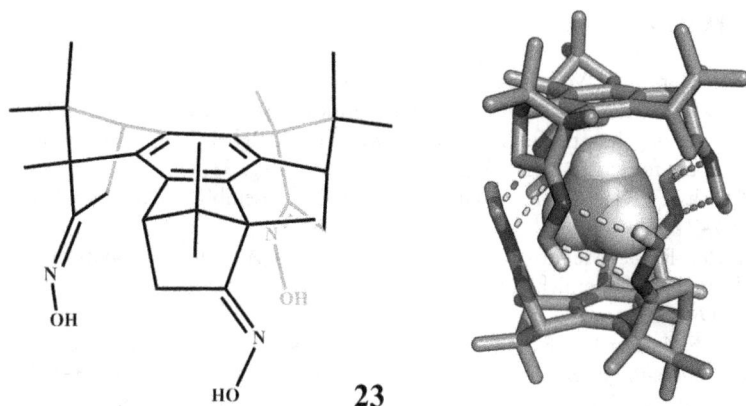

Figure 1.27   *Left*: A self-complementary trioxime. *Right*: The dimeric capsule with CH_4 guest.

C_{76}, C_{78}, C_{82} and C_{84}. The concave surface of CTV is an appropriate complement for the convex surfaces of fullerenes; Atwood had earlier recognized this and named their complexes with C_{60} "ball and socket" structures.[52]

At the other extreme — in terms of volume — Scarso has synthesized a small capsule which hydrogen bonds through oxime functions (Figure 1.27).[53] The curvature was provided by the readily available (+)-syn-benzotriborneol,[54] which has recognition properties of its own. Oxidation and then oximation gave the module **23** with (the heretofore unrecognized) self-complementary hydrogen bond components presented by the oximes.[55]

Atmospheric gases are taken up in the dimeric capsule, and broadening of the NMR signals in $CDCl_3$ or C_6H_6 with added O_2 suggest that the oxygen molecule is a guest — an unprecedentd observation. The direct observation of encapsulated CH_4 was possible by ^{1}H NMR at lowered temperature; the signal appears at -2.64 ppm. Under ambient conditions the gas exchanges in and out rapidly on the NMR timescale and an averaged signal for the environments is seen. While many examples of gases in covalent capsules exist,[56] supramolecular sensing of gases in hydrogen-bonded capsules is rare.

References

1. (a) Galán A, de Mendoza J, Toiron C, Bruix M, Deslongchamps G, Rebek Jr J. (1991) A synthetic receptor for dinucleotides. *J Am Chem Soc* **113**, 9424–9425; (b) Deslongchamps G, Galán A, de Mendoza J, Rebek Jr J. (1992) A synthetic receptor for cyclic AMP. *Angew Chem Int Ed Engl* **31**, 61–63.

2. Rebek Jr J. (1996) Molecular recognition and assembly. *Acta Chem Scand* **50**, 707–716.

3. Wyler R, de Mendoza J, Rebek Jr J. (1993) A synthetic cavity assembles through self-complementary hydrogen bonds. *Angew Chem Int Ed Engl* **32**, 1699–1701.

4. Hof F, Palmer LC, Rebek Jr J. (2001) Synthesis and self-assembly of the tennis ball and subsequent encapsulation of methane. *J Chem Ed* **78**, 1519–1521.

5. Branda N, Wyler R, Rebek Jr J. (1994) Encapsulation of methane and other small molecules in a self-assembling superstructure. *Science* **263**, 1222–1223.

6. Valdés C, Spitz UP, Toledo L, Kubik S, Rebek Jr J. (1995) Synthesis and self-assembly of pseudo-spherical homo- and heterodimeric capsules. *J Am Chem Soc* **117**, 12733–12745.

7. Chapman KT, Still WC. (1989) A remarkable effect of solvent size on the stability of a molecular complex. *J Am Chem Soc* **I11**, 3075–3077.

8. (a) Quan MLC, Cram DJ. (1991) Constrictive binding of large guests by a hemicarcerand containing four portals. *J Am Chem Soc* **113**, 2754–2755; (b) Robbins TA, Knobler CB, Bellew DR, Cram DJ. (1994) A highly adaptive and strongly binding hemicarcerand. *J Am Chem Soc* **116**, 7717–7727.

9. Rebek Jr J. (1987) Model studies in molecular recognition. *Science* **235**, 1478–1484.

10. Adrian Jr JC, Wilcox CS. (1989) Chemistry of synthetic receptors and functional group arrays. 10. Orderly functional group dyads. Recognition of biotin and adenine derivatives by a new synthetic host. *J Am Chem Soc* **111**, 8055–8057.

11. Zimmerman SC, Wu WA. (1989) Rigid molecular tweezers with an active site carboxylic acid: exceptionally efficient receptor for adenine in an organic solvent. *J Am Chem Soc* **111**, 8054–8055.

12. Sanderson PEJ, Kilburn JD, Still WC. (1989) Enantioselective complexation of simple amides by a C2 host molecule. *J Am Chem Soc* **111**, 8314–8315.

13. (a) Dixon RP, Geib SJ, Hamilton AD. (1992) Molecular recognition: bis-acylguanidiniums provide a simple family of receptors for phosphodiesters. *J Am Chem Soc* **114**, 365–366; (b) Paliwal S, Geib S, Wilcox CS. (1994) Molecular torsion balance for weak molecular recognition forces. Effects of "tilted-T" edge-to-face aromatic interactions on conformational selection and solid-state structure. *J Am Chem Soc* **116**, 4497–4498.

14. Cram DJ, Karbach S, Kim YH, *et al.* (1985) Shell closure of 2 cavitands forms carcerand complexes with components of the medium as permanent guests. *J Am Chem Soc* **107**, 2575–2576.

15. Canceill J, Cesario M, Collet A, *et al.* (1985) A new bis-cyclotribenzyl cavitand capable of selective inclusion of neutral molecules in solution — crystal-structure of its CH_2Cl_2 cavitate. *J Chem Soc Chem Comm* 6, 361–363.

16. Branda NR, Grotzfeld RM, Valdés C, Rebek Jr J. (1995) Control of self-assembly and reversible encapsulation of xenon in a self-assembling dimer by acid-base chemistry. *J Am Chem Soc* 117, 85–88.

17. Valdés C, Spitz UP, Toledo L, Kubik S, Rebek Jr J. (1995) Synthesis and self-assembly 1of pseudo-spherical homo- and heterodimeric capsules. *J Am Chem Soc* 117, 12733–12745.

18. Pitera J, Kollman P. (1998) Designing an optimum guest for a host using multi-molecule free energy calculations: predicting the best ligand for Rebek's "tennis ball." *J Am Chem Soc* 120, 7557–7567.

19. Valdés C, Toledo L, Spitz UP, Rebek Jr J. (1996) Structure and selectivity of a small dimeric encapsulating assembly. *Chem Eur J* 2, 989–991.

20. Timmerman P, Verboom W, Van Veggel FCJM, *et al.* (1994) *Angew Chem Int Ed Engl* 33, 1292–1295.

21. Szabo T, Hilmersson G, Rebek Jr J. (1998) Dynamics of assembly and guest exchange in the tennis ball. *J Am Chem Soc* 120, 6193–6194.

22. Hammes GG, Park AC. (1968) Kinetic studies of hydrogen bonding. 1-cyclohexyluracil and 9-ethyladenine. *J Am Chem Soc* 90, 4151–4157.

23. Jeong KS, Rebek Jr J. (1988) Molecular recognition: hydrogen bonding and aromatic stacking converge to bind cytosine derivatives. *J Am Chem Soc* 110, 3327–3328.

24. Jorgensen WL, Pranata J. (1990) Importance of secondary interaction in triply hydrogen bonded complexes: guanine–cytosine vs uracil-2,6-diaminopyridine. *J Am Chem Soc* 112, 2008–2010.

25. Nishio M, Hirota M, Umezawa Y. (1998) In *The CH/π Interaction: Evidence, Nature, and Consequences. Stereochemistry of Organic Compounds.* Wiley, New York.

26. Grotzfeld R, Branda N, Rebek Jr J. (1996) Reversible encapsulation of disc-shaped guests by a synthetic, self-assembled host. *Science* 271, 487–489.

27. O'Leary BM, Grotzfeld RM, Rebek Jr J. (1997) Ring inversion dynamics of encapsulated cyclohexane. *J Am Chem Soc* 119, 11701–11702.

28. A phrase coined by D. J. Cram as the title of his lecture at the C. David Gutsche Symposium, Washington University, St. Louis, Missouri, May 5, 1990.

29. Kang J, Meissner RS, Wyler R, de Mendoza J, Rebek Jr J. (2000) Development of synthetic self-assembling molecular capsule: from flexible spacer to rigid spacer. *Bull Korean Chem Soc* 21, 221–224.

30. Meissner R, Rebek Jr J, de Mendoza J. (1995) Autoencapsulation through inter-molecular forces: a synthetic self-assembling spherical complex. *Science* 270, 1485–1488.

31. Kang J, Rebek Jr J. (1997) Acceleration of a Diels–Alder reaction by a self-assembled molecular capsule. *Nature* 385, 50–52.

32. Kang J, Rebek Jr J. (1996) Entropically-driven binding in a self-assembling molecular capsule. *Nature* 382, 239–241.

33. Santamaria J, Martîn T, Hilmersson G, Craig SL, Rebek Jr J. (1999) Guest exchange in an encapsulation complex: a supramolecular substitution reaction. *Proc Natl Acad Sci USA* **96**, 8344–8347.
34. Wang X, Houk KN. (1999) Gating and entropy in guest exchange by Rebek's sportsballs. Theoretical studies of one-door, side-door, and back-door gating. *Org Lett* **1**, 591–594.
35. Garcias X, Toledo LM, Rebek Jr J. (1995) Synthesis and solid state structure of an unsymmetrical triurea. *Tetrahedron Lett* 8535–8538.
36. Martin T, Obst U, Rebek Jr J. (1998) Hydrogen-bonding preferences and the filling of space provide information for molecular assembly and encapsulation. *Science* **281**, 1842–1845.
37. Johnson DW, Hof F, Iovine PM, Nuckolls C, Rebek Jr J. (2002) Solid state and solution studies of a tetrameric capsule and its guests. *Angew Chem Int Ed Engl* **41**, 3793–3796.
38. Metzger A, Lynch VM, Anslyn EV. (1997) A synthetic receptor selective for citrate. *Angew Chem Int Ed Engl* **36**, 862–865.
39. Szabo T, O'Leary B, Rebek Jr J. (1999) Self-assembling sieves. *Angew Chem Int Ed Engl* **37**, 3410–3413.
40. Lützen A, Renslo AR, Schalley CA, O'Leary BM, Rebek Jr J. (1999) Encapsulation of ion-molecule complexes: second-sphere supramolecular chemistry. *J Am Chem Soc* **121**, 7455–7456.
41. Kamitori S, Hirotsu K, Higuchi T. (1987) Crystal and molecular structures of double macrocyclic inclusion complexes composed of cyclodextrins, crown ethers, and cations. *J Am Chem Soc* **109**, 2409–2414.
42. Vögtle F, Müller WM. (1979) Complexes of γ-cyclodextrin with crown ethers, cryptands, coronates, and cryptates. *Angew Chem Int Ed Engl* **18**, 623–624.
43. Chapman RG, Sherman JC. (1995) Study of templation and molecular encapsulation using highly stable and guest-selective self-assembling structures. *J Am Chem Soc* **117**, 9081–9082.
44. Chapman RG, Olovsson G, Trotter J, Sherman JC. (1998) Crystal structure and thermodynamics of reversible molecular capsules. *J Am Chem Soc* **120**, 6252–6260. See also: Dumitrescu D, Dumitru F, Legrand YM, *et al.* (2015). New "pyrene box" cages for adaptive guest conformations. *Org Lett* **17**(9): 2178–2181.
45. Mansikkamäki H, Nissinen M, Schalley CA, Rissanen K. (2003) Self-assembling resorcinarene capsules: solid and gas phase studies on encapsulation of small alkyl ammonium cations. *New J Chem* **27**, 88–97.
46. Kobayashi K, Shirasaka T, Yamaguchi K, *et al.* (2000) Molecular capsule constructed by multiple hydrogen bonds: self-assembly of cavitand tetracarboxylic acid with 2-aminopyrimidine. *Chem Commun* 41–42.
47. Kobayashi K, Ishii K, Sakamoto S, *et al.* (2003) Guest-induced assembly of tetracarboxyl-cavitand and tetra(3-pyridyl)-cavitand into a heterodimeric capsule via hydrogen bonds and CH–halogen and/or CH–π interaction: control of the orientation of the encapsulated guest. *J Am Chem Soc* **125**, 10615–10624.

48. Kobayashi K, Yamanaka M. (2014) Self-assembled capsules based on tetrafunctionalized calix[4]resorcinarene cavitands. *Chem Soc Rev* DOI: 10.1039/c4cs00153b.

49. Park YS, Paek K. (2008) Hydrazide as a new hydrogen-bonding motif for resorcin[4]arene-based molecular capsules. *Org Lett* **10**, 4867–4870.

50. Aakeroy CB, Rajbanshi A, Desper J. (2011) Hydrogen-bond driven assembly of a molecular capsule facilitated by supramolecular chelation. *Chem Commun* **47**, 11411–11413.

51. Huerta E, Metselaar GA, Fragoso A, *et al.* (2007) Selective binding and easy separation of C70 by nanoscale self-assembled capsules. *Angew Chem Int Ed* **46**, 202–205.

52. Steed JW, Junk PC, Atwood JL. (1994) Ball and socket nanostructures: new supramolecular chemistry based on cyclotriveratrylene. *J Am Chem Soc* **116**, 10346–10347.

53. Scarso A, Pellizzaro L, De Lucchi O, *et al.* (2007) Gas hosting in enantiopure self-assembled oximes. *Angew. Chem Int Ed* **46**, 4972–4975.

54. Fabris F, Pellizzaro L, Zonta C, De Lucchi O. (2007) A novel C3-symmetric triol as chiral receptor for ammonium ions. *Eur J Org Chem* 283–291.

55. Bruton EA, Brammer L, Pigge FC, *et al.* (2003) Hydrogen bond patterns in aromatic and aliphatic dioximes. *New J Chem* **27**, 1084–1094.

56. Rudkevich DM. (2007) Progress in supramolecular chemistry of gases. *Eur J Org Chem* 3255–3270.

CHAPTER 2

Calixarene Capsules

At this time (1995–1996) Ken Shimizu, surrounded by the encapsulation studies going on in the research group, decided to invent his own capsule, **1.1** (Figure 2.1). The concave surface he chose was based on the calixarenes developed by Gutsche,[1] and the hydrogen bond motif expected to hold the two calixarene hemispheres together was inspired by Peggy Etter's studies of urea hydrogen-bonding patterns in the solid state.[2] Unbeknownst to us, two other research groups were arriving at the calixarene capsules: those of Volker Böhmer in Mainz and David Reinhoudt[3] in Enschede. In Europe, calixarenes were already much-admired as platforms for crown ethers and had been well developed for the selective sequestration of Cs+ in their "pinched cone" conformations.[4] These compounds are so reliable that they are still in use today in the American nuclear waste industry.

Shimizu sedulously fixed the cone conformation using benzyl groups on the lower periphery and built the hydrogen-bonding seam through upper rim amination and then reaction with isocyanates, as shown in the figure.[5] The dimeric assembly has a sense of chirality, since the head-to-tail ureas can be clockwise or counterclockwise. This asymmetry gives the CH_2 groups of the benzyl ethers diastereotopic status. This was suggested to us by Böhmer[6] and is apparent in the NMR spectrum shown.

The upfield shifts in the NMR spectrum of the guests were, of course, the other piece of evidence that encapsulation had occurred. Again, separate signals were seen for guests inside and outside (slow exchange). Shortly thereafter, Böhmer obtained crystallographic evidence for encapsulation[7] and he continued to make increasingly exotic arrays of calixarene capsules by combining the dynamism of reversible systems with the permanence of mechanically bonded ones.[8]

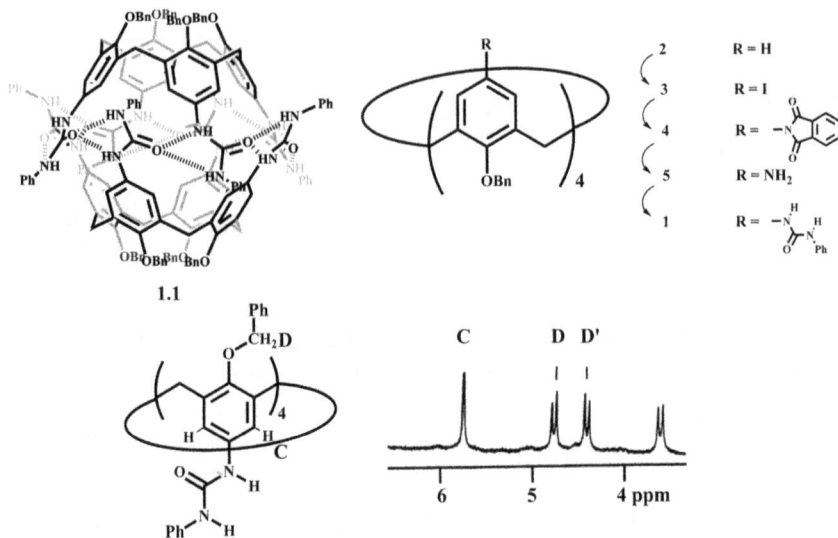

Figure 2.1 *Top*: Structure of the capsule **1.1**, showing the hydrogen-bonded ureas and the synthetic sequence. *Bottom*: A partial NMR spectrum showing the diastereotopic signals (D) for the benzyl groups. [From *PNAS* **92**, 12403–12407. Copyright 1995, National Academy of Sciences of the United States of America.]

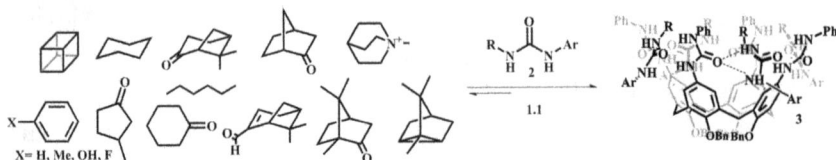

Figure 2.2 Typical guests of the calixarene capsule **1.1** and its "deconstruction" by the addition of competitive ureas **2**.

Shimizu and Blake Hamann probed the space inside the capsule (two square pyramids rotated at 45° from each other) by taking up groups of various sizes (Figure 2.2) and shapes,[9] and found that even short alkanes such as pentane can be encapsulated.[10] Again, the solvent size proved critical; in this case *p*-xylene-d_{10} was not encapsulated and was sufficiently free of smaller impurities to allow intended guests access. For the first time, we could see that certain guests had preferred orientations in the capsule. Specifically, the chemical shifts of the *ortho, meta* and *para* protons of C_6H_5F indicated its positioning in such a way that the polar C–F bond is directed toward the seam of hydrogen bonds and the *ortho* and *meta* hydrogens are directed toward the aromatic panels. The orientation preferences induced by the lining of the capsule were subsequently recruited to fix orientations in other, nonspherical capsules — about which more later.

We also encountered the phenomenon of "deconstruction" for the first time. The addition of simple aryl ureas **2** to the capsular assemblies **1.1** released the guest; the simple ureas were inserted into the seam and disrupted it to give **3**. But "reconstruction" was also a feature of this type of capsule. Ron Castellano and Dmitry Rudkevich found that a calixarene without a conformational preference but with the tetraurea rim could be coaxed into the cone shape and the capsular host by simply exposing it to a good guest.[11] Likewise, hybridization with conformationally "fixed" tetraurea calixarenes occurred to give capsules.

In a collaboration with Yoram Cohen of Tel Aviv, the Böhmer group used diffusion-ordered spectroscopy (DOSY) to show that this capsule and its contents (benzene) diffused as a single entity.[12] This use of DOSY was an innovation and represented a huge and enabling step forward in supramolecular methodology; it has been widely applicable[13] and we have used it as well.

A second innovation that occurred at that time allowed capsules to be characterized in the gas phase. Solvents generally used for protonation of the assemblies for mass spectrometry are protic and disrupt the hydrogen bonds of the capsules. But Christoph Schalley, Ron Castellano, Marcus Brody and Dmitry Rudkevich joined my colleague Professor Gary Siuzdak to apply cation–π interactions to the problems. As previously mentioned, the interior lining of the capsules features many polarizable π bonds that show high affinity for cations. The team realized that cations could be *simultaneously* guests and labels for soft ionization methods such as electrospray.[14] They determined that the N-methylquinuclidinium ion, for example, was a superior guest for the calixarene capsules, in solution and in the gas phase.

But the most fruitful of these hybridizations involved calixarenes with tetrasulfonylamides on the upper rim. Castellano used

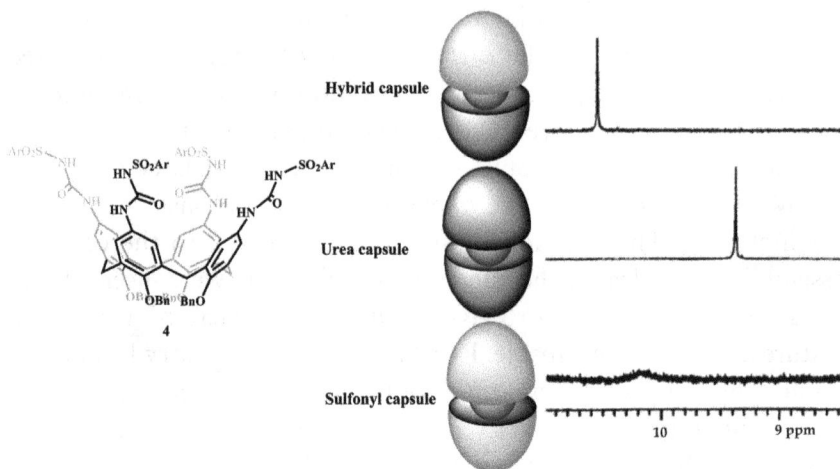

Figure 2.3   The sulfonylurea module **4** also dimerizes to form the sulfonyl capsule. When the sulfonyl and urea capsules are mixed, only a hybrid capsule can be detected in the NMR spectrum.

p-toluenesulfonyl isocyanate in the final step of the synthesis and found that these functionalized calixarenes **4** (Figure 2.3) would also form dimers (sulfonyl capsules).

Unexpectedly, a mixture of the arylurea capsules with the sulfonylurea capsules led to *exclusive hybridization to heterodimer capsules* (Figure 2.3). Once again, self-sorting was not a feature of these self-assembly processes. Presumably, the better hydrogen bond donors (sulfonylureas) prefer to match up with the better acceptors (the arylureas) to reliably give capsules with two different hemispheres.[15]

The extensive manipulations of calixarenes in other laboratories led to reliable means of monofunctionalizing either the upper or the lower rim. It seemed obvious that two calixarenes could be covalently attached as in structure **5** (Figure 2.4): on the upper rim a flexible bifunctional group might orient the ureas toward each other and permit the formation of a covalently linked unimolecular capsule **6**, a pair of capsules **7** or a polymeric array **8**.

But covalent connections on the lower rim as in **9** could orient the tetraureas away from each other and result in polymerization **10** when capsules formed. Both types of compounds were eventually

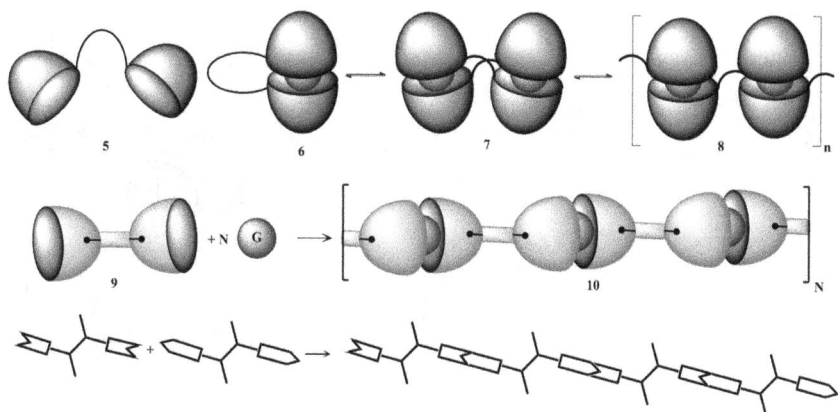

Figure 2.4 Types of self-assembly in structure forms. *Top*: Intramolecular and intermolecular assembly of self-complementary monomers. *Center*: Polymerization of a self-complementary calixarene. *Bottom*: Lehn's noncovalent polymerization of two complementary components.

synthesized and characterized, but the polymeric system came to hand first.

Castellano and Rudkevich covalently attached two lower rim functionalized calixarenes **10** to the ends of a simple *p*-substituted benzene spacer **11**. With proper attention to the length and rigidity of the spacer, they could guarantee that a unimolecular capsule could not form from the module **12**. Instead, when the conditions were appropriate, the capsules formed as notional beads on a string — as polymers. We named them "polycaps"[16] (Figure 2.5).

At that time, the only supramolecular polymers known had been introduced by Lehn in 1990.[17] They involved hydrogen-bonding between two bifunctional and complementary partners and the work inspired enormous activity in materials science. The self-complementary calixarenes allowed the examination of completely surrounded guest species in increasingly ordered phases, and we were anxious to take advantage of this unique property of polycaps.

By that time, we had moved to Scripps from MIT, but retained our ties to that institution by recruiting Professor George Benedek and Aleksey Lomakin[18] to use quasielastic light scattering to show

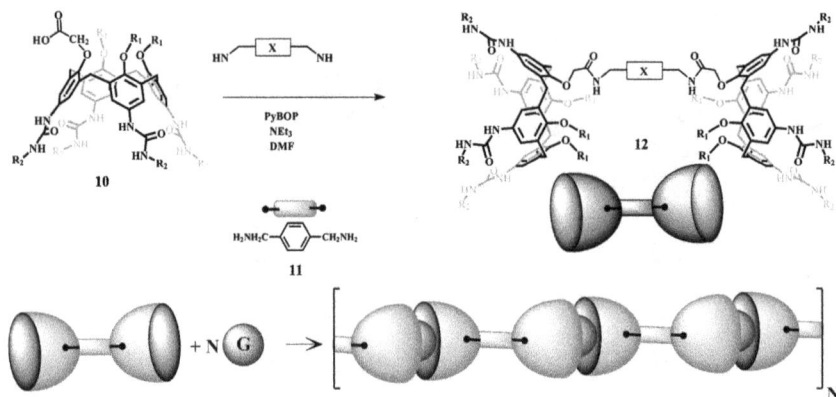

Figure 2.5 *Top*: Structure and synthesis of a "polycap" module. *Bottom*: Cartoon of self-assembly in the presence of guests G.

the degree of polymerization of the polycaps. This provided a spectroscopic method beyond the scope of NMR that could be applied to this range of supramolecular chemistry. Size determinations involving equilibrium polymerization of such "living" polymers in dilute solution have two requirements: the dynamics must be appropriate on the timescale of chain diffusion, and the energetics must be appropriate for giving assemblies of sufficient size. The measurements of the degree of polymerization were 10–60 in the millimolar concentration range (in $CHCl_3$). The molecular weights are 31,000 to 186,000 and the polycap chains preserve their direction through about four monomer units. The derived association constants, $\sim 2 \times 10^6\,M^{-1}$, are in agreement with the values obtained by fluorescence studies described below.

As we were unfamiliar with the techniques of polymer characterization, Holger Eichhorn in Timothy Swager's lab at MIT, Andrew Lovinger at the Bell Labs, and Ross Clark of Kelco Biopolymers (San Diego) were also enlisted as collaborators. A number of polymeric capsules were examined, including cross-linked structures with molecules like difluorobenzene and nopinone as guests.[19] While this is not the place to examine the behavior in detail (the details are published elsewhere),[20] some unique properties are worth mentioning.

There were some expected differences between reversible polymers and covalent ones. Covalent polymers generally move via the "reptation" mechanism, where chains work their way through a virtual tube to escape entanglements. This process is slow and concentration-dependent, but hydrogen-bonded polymers have an alternative: the monomers can simply dissociate and rearrange out of the entanglements. Rheological studies showed this to be the case.

Also as expected, polycaps are dramatically sensitive to chemical changes. Competitive solvents disrupt the assembly; for example, merely 5% added methanol decreases the viscosity by two orders of magnitude. Removal of the methanol by heating for a few minutes returns the viscosity to its initial value.

We also observed a liquid crystalline behavior from suitable derivatives of the polycaps which ordered themselves into a nematic phase. The X-ray diffraction pattern of the liquid crystalline samples showed two main peaks at 2.4 nm and 1.6 nm, which are attributed to the self-organization of the mesogenic phase (Figure 2.6). The 2.4 nm spacing matches the repeat unit along the polymer chain, while

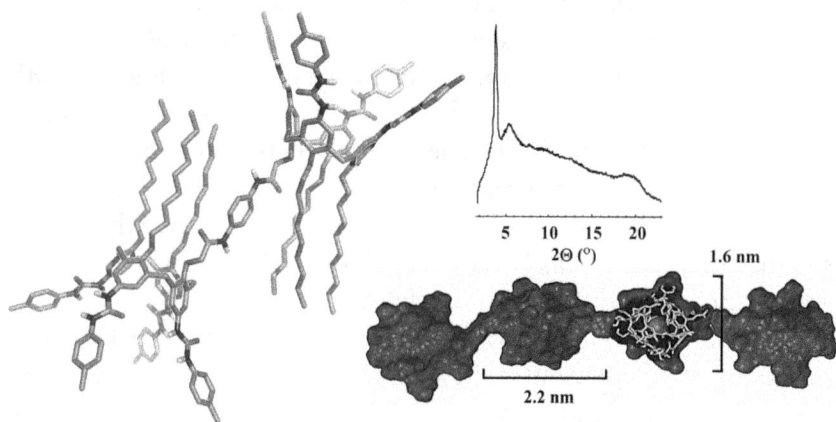

Figure 2.6 *Left*: Attachment of long aliphatic chains to **12** gave liquid crystalline polycaps. *Right*: The low angle X-ray diffraction pattern shows peaks corresponding to the beads-on-a-string model. [Reprinted with permission from *Angew Chem Int Ed Eng* **38**, 2603–2606. Copyright 1999, Wiley-VCH, Weinheim.]

Figure 2.7 *Left*: The fiber of a cross-linked polycap drawn from solution. *Right*: The polycap fiber shows tensile strength comparable to that of nylon. [Reprinted with permission from *Angew Chem Int Ed Eng* **38**, 2603–2606. Copyright 1999, Wiley-VCH, Weinheim.]

the 1.6 nm, spacing corresponds to the distance between neighboring polymer chains.

We were able to pull fiber structures from the polymer liquid crystals. These well-ordered fibers can be pulled to centimeters in length and their width (about 6 micrometers) was remarkably similar to those that resulted from shearing. They have appreciable strength; Figure 2.7 shows the fiber supporting the weight of tape and an NMR tube cap. The picture shows Steven Craig's method of evaluation. These polycap fibers have a tensile strength on the order of 10^8 Pa, as measured by the load at break, which is approximately the same as for commercial nylon fibers. The polycap fibers also have the advantage that they can be easily redissolved and redrawn, while covalent polymer fibers are difficult to recycle.

The reversible polymer was strong enough to withstand disruptive conditions of stresses, but some of the most dramatic effects had to do with three-dimensional polymer networks. Such systems are commonplace in biology, particularly among polysaccharides and protein fibers. We used PAMAM **13**, much admired as a "zeroth"-generation dendrimer, as the tetramine and covalently connected it to the calixarene **10** to give the cross-linker module **14** in one step (Figure 2.8).

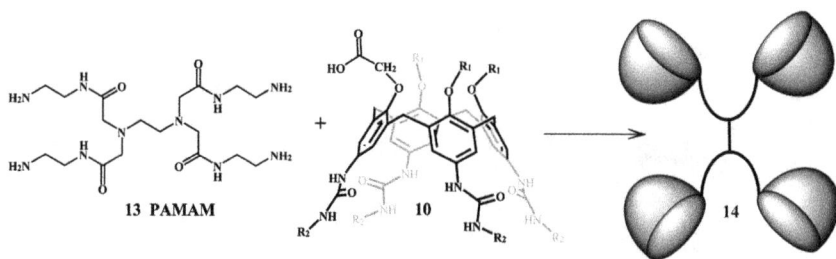

Figure 2.8 Synthesis of a module for cross-linked polycaps.

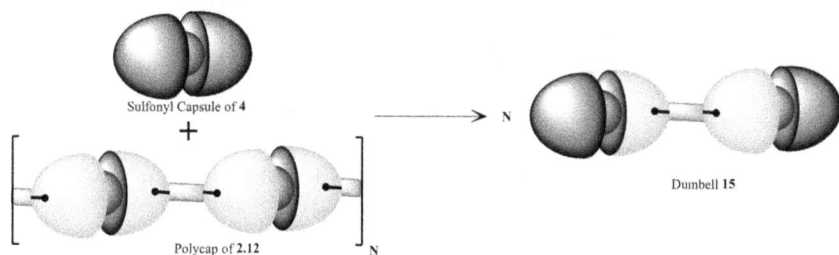

Figure 2.9 Deconstruction of a polycap to a dumbbell. The exclusive hybridization of urea and sulfonylurea calixarene modules drives the disproportionation.

The assembled polycap network had a significant mechanical integrity in *o*-dichlorobenzene solution and a greater elastic component than the simple linear polycaps. A gel formed by 0.5% of the cross-linked polymer in this solvent was rigid on short timescales, but flowed on longer timescales.

What are at the ends of the polycaps? This question has several answers, such as unpaired ureas or no ends, as in macrocycles (in the largest sense). We performed deliberate end-capping experiments using simple dimeric capsules. When added in excess to the polycaps, the polymer broke down to a dumbbell-shaped assembly **15** comprising two guests and three host particles (Figure 2.9). This assembly showed sharply resolved NMR signals for all of the expected resonances.

The reliable means of monofunctionalizing the lower rim was used to covalently attach the calixarenes to various spacers. The exclusive hybridization allowed several discreet (versus polymeric)

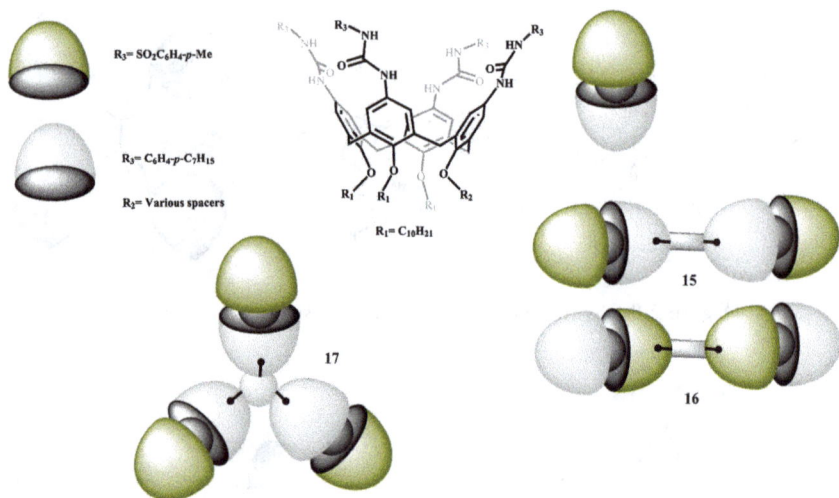

Figure 2.10 The exclusive hybridization of urea and sulfonylurea calixarene modules allows the synthesis of discrete multicomponent assemblies.

nanoscale assemblies beyond the dumbells **15** and **16** to be predictably constructed.[21] For example, the spacer tren gave **17** a seven-component assembly (Figure 2.10).

The calixarene capsules were also our first experience with a chiral capsule lining. As previously mentioned, the cycloenantiomerism of the head-to-tail ureas imparts an asymmetry to each capsule. When the two hemispheres are identical, the clockwise and counterclockwise arrangements are interconverted by just turning the capsule upside down. But if the two hemispheres are different, chiral (but racemic) capsules are formed. This was first encountered by Böhmer, who showed that two (octaurea) symmetrical capsules of this sort disproportionated to give a statistical distribution of assemblies that included the heterodimers.[22] However, the steric environment of the capsule is not much affected by this, as the lining of the capsule is relatively smooth. In macroscopic terms, it is as though a chiral painting was placed on the walls of an otherwise symmetrical capsule. The urea/sulfonyl urea heterodimeric capsules are also racemic, but again their spaces are not very chiral, and we will explore this more fully in the chapter devoted to chiral spaces (Chapter 6). For the time being, the chiral nature of the space versus the chiral nature of the magnetic environment is a recurrent issue in many capsules, but consider the analogous case with cyclodextrins. Despite their large number of asymmetric centers, typically 25–30, they are not particularly good at enantioselection; their interiors are rather smooth and they have very high symmetries. This is also the case with the capsules in hand.[23]

Useful as NMR is for structure determination in solution, its operation at millimolar concentrations prevents the study of rates and equilibria of capsule formation because these processes occur too rapidly. The large number of hydrogen bonds (up to 16 holding the calixarene capsules together) require more dilute (micro- to nanomolar) concentrations for equilibria to be evaluated, and led us to apply fluorescence resonance energy transfer (FRET) methods for studies of assembly/dissociation processes in real-time.

While FRET was already widely used in biological systems,[24] we were among the earliest to apply it to synthetic self-assemblies, and certainly the first — if not the only ones — to apply it to capsules. The plan was simple: attach donor (D) and acceptor (A) dyes to the respective calixarenes and monitor the energy transfer as the two halves come together and the assembly of the capsules takes place. The monofunctional lower rims were ideally primed to attach fluorescent dyes to the ureas **10** and sulfonylureas **18** (Figure 2.11). The necessary donor emission/acceptor absorption overlap was found in the spacer-modified coumarin 2 as D (structure **19**) and the coumarin 343 as A (structure **24**).

This pair constitutes a convenient Stokes shift of approximately 100 nanometers. An ethylenediamine spacer **21** was used and the appropriate calixarenes that were synthesized comprised the sulfonylurea donor **2.20** (cartoon SD), the urea donor **22** (cartoon UD) and the urea acceptor **23** (cartoon UA). Control experiments with dye derivatives established that intermolecular FRET between the dyes was negligible below 500 nM concentrations in *p*-xylene. The results obtained for solvent guests by Ron Castellano, Stephen Craig and Colin Nuckolls are summarized in Figure 2.12 and Table 2.1.[25]

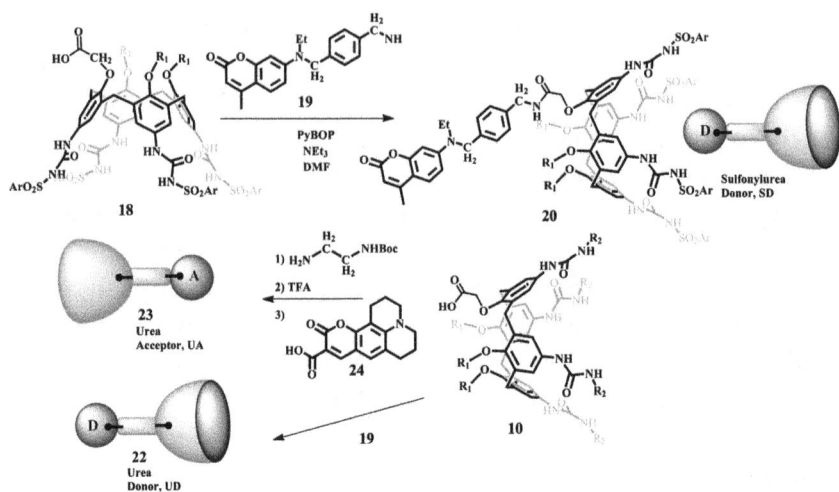

Figure 2.11 Synthesis of calixarenes labeled with fluorophores enabled FRET studies of capsule assembly in real-time.

These values are compatible with those from light scattering in these solvents but, since the capsules assemble only when a guest is present, it was possible to use this system as a sensor for small molecule guests when the solvent is not a guest. The guest 3-methylcyclopentanone could be detected this way in p-xylene with the capsule at 150 nM concentrations.

Eventually, Marcus Brody and Chris Schalley attached two calixarenes covalently by their upper rims as **25** (Figure 2.13). A hexyl group acted as a flexible spacer and the formation of a unimolecular capsule **26** was evident with the guest, N-methyl-quinuclidinium.[26] The compound showed no tendency toward dimerization or oligomerization but did form the heterodimeric *bis*-capsule **27** with the simple sulfonylurea calixarene **4**. The tethered **25** straddles the line between covalent and reversible capsules, and its attachment to an insoluble solid support could constitute a polymer-bound sensor or sequestering agent for small molecules.

In an ingenious application, Rudkevich adapted the polymeric calixarene capsules for sensing of CO_2. In this case the gas was not itself encapsulated but reacted with primary amine sites planted along the

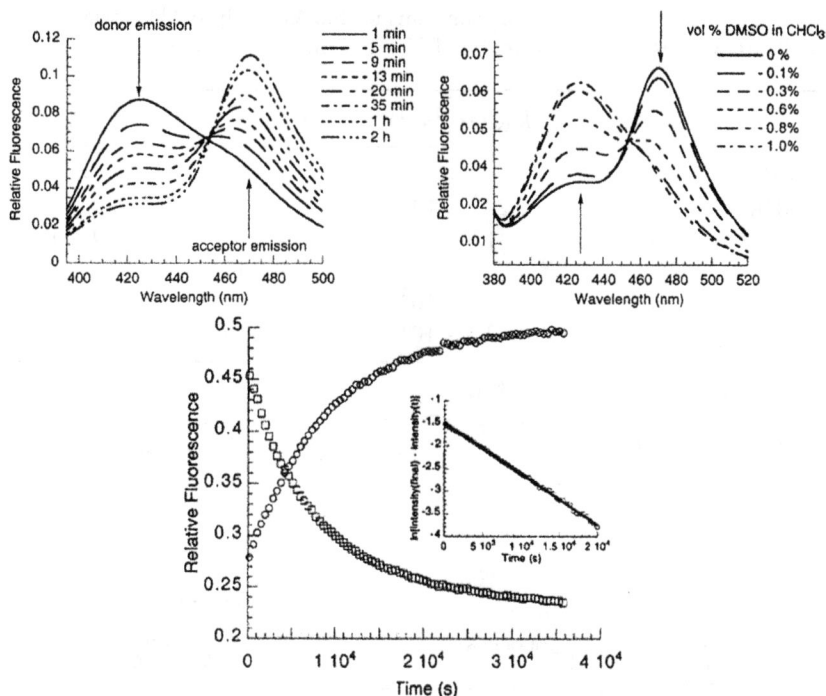

Figure 2.12 *Top*: FRET as a function of time in the assembly of the capsule (*left*); the effects of DMSO in dissipation of the capsule (*right*). *Bottom*: Real-time rate measurements. (Reprinted with permission from *J Am Chem Soc* 122, 7876–7882. Copyright 2000, American Chemical Society.)

polymer backbone.[27] The resulting carbamates made salt bridges with other ammonium sites that caused polymer cross-linking and precipitation. However, many examples of gas sensing by other forms of encapsulation have been reported.[28]

Because of their size limitations, the expansion of calixarenes to larger cavities is desirable. Attempts to increase their sizes in another dimension has not yet yielded promising results. The calix[5]pentaurea does not assemble[29] and the bigger calix[6]case was examined by de Mendoza *et al.*[30] This involved the methodical functionalization of a calix[6]arene (Figure 2.14) with three *t*-butyl groups and three ureas on the upper rim, as in **28**. Assembly and encapsulation of solvent-sized molecules, even water,[31] was observed as in **29**,

Table 2.1 Rate and Association Constants for the Assembly and Dissociation of Calixarene Capsules as Measured by FRET at Concentrations of 0.05–0.5 μM.

	Solvent	$K_{ass}(M^{-1}s^{-1})^a$	$K_{diss}(s^{-1})^a$	$K_A(M^{-1})^a$
SD.SD	CHCl$_3$	—^{b}	—^{b}	—^{b}
SD.UA	CHCl$_3$	$2.7^b \times 10^3$	2.3×10^{-5}	1.2×10^8
UA.UD	CHCl$_3$	1.5×10^4	6.4×10^{-3}	2.4×10^6
SD.SD	C$_6$H$_6$	<50	5.4×10^{-5}	$<10^6$
SD.UA	C$_6$H$_6$	3×10^3	5×10^{-6}	6×10^8
UA.UD	C$_6$H$_6$	2.5×10^4	6×10^{-4}	4×10^7

^{a}Uncertainties in K_{diss} are ±15% and in K_{ass} and K_A are ±40%.
^{b}Compound **20** (**SD**) does not appear to form a well-defined dimer in CHCl$_3$ by NMR, and its behavior is consistent with a species that is only monomeric at these submicromolar concentrations.

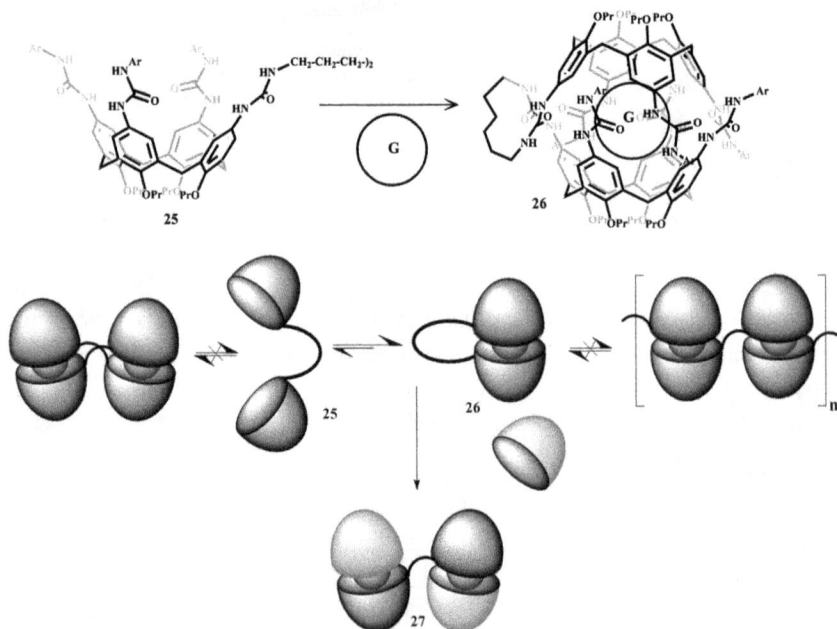

Figure 2.13 The unimolecular urea capsule hybridizes with the sulfonylurea calixarene modules to assemble conjoined capsules **27**.

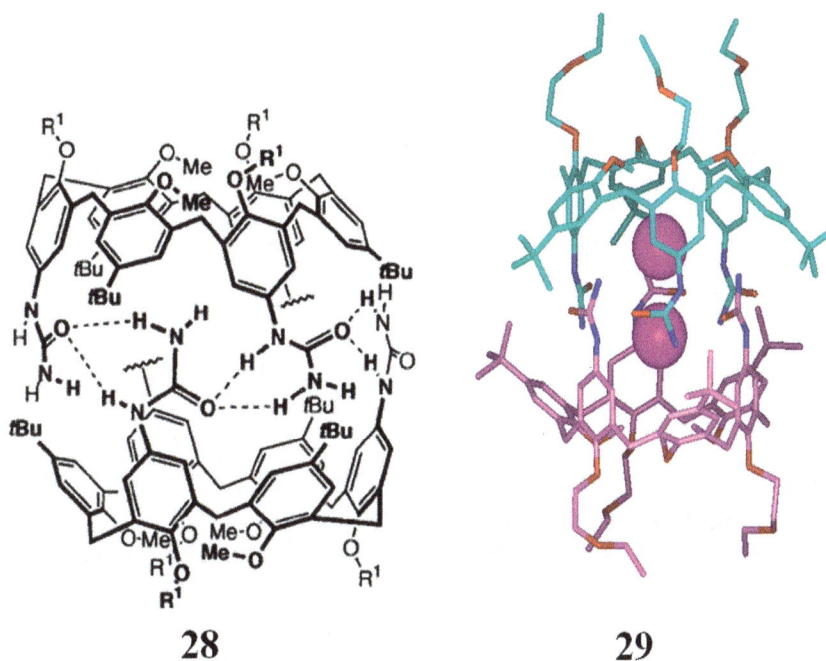

28   **29**

Figure 2.14 *Left*: Schematic of a calix[6]arene dimer with alternating ureas and *t*-butyl groups on the upper rim. *Right*: X-ray structure of the dimer with disordered solvent guests. [Reprinted with permission from *Chem Eur J* 6(1), 73–80. Copyright 2000, Wiley-VCH, Weinheim.]

but larger guests were not taken up. Modeling indicated an effective volume comparable to the calix[4]arene dimers.

Calixpyrroles

Another source of capsules with the added charm of inwardly directed hydrogen bond donors are the calixpyrroles. These compounds **2.30** (Figure 2.15) are also known as porphyrinogens. They were exhumed by Sessler,[32] who resurrected them and recognized their huge potential for anion binding. Uncannily, calixpyrroles and their-not-so-remotely-related recorcinarene cousins[33] were invented by the same German chemist, Baeyer, more than a century ago.[34] A variety of conformations are available to these macrocycles in nonpolar media

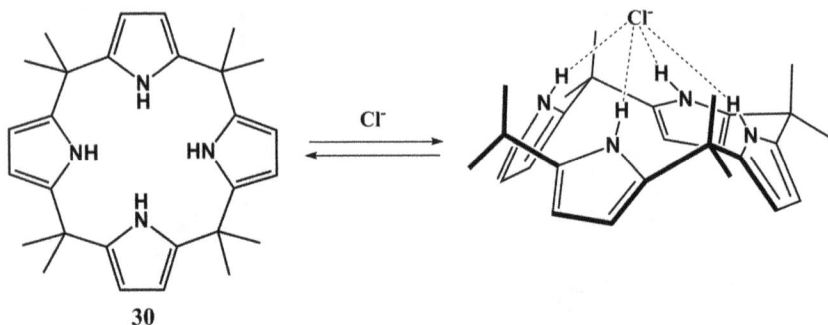

30

Figure 2.15 Calixpyrroles **30** show convergent N–H hydrogen bond donors in the presence of acceptor anions.

involving different orientations of their pyrrole rings. But in the presence of anions or other polar H-bond acceptors, all four pyrrole rings converge their N–H bonds to act as cooperative and forceful hydrogen bond donors.

Deeper versions[35] with fixed wall hydrogen-bonding sites **31** (Figure 2.16) can form dimers and take up molecules such as ethanol in capsule **32**. Ballester prepared the analog of the calixarene tetraurea **33** and showed that it forms the same type of capsule **34** in the presence of guests. The added value of introverted N–H bonds in the capsule is clear with N-oxide guest **35**. The capsule's hydrogen bond donors find their complements in the correct orientation and distance provided by the guest. An unusual "sorting" process involving calixarene and calixpyrrole tetraureas produces a hybrid dimeric capsule **36**. The self-sorting is not between homo- and heterodimeric structures; rather, it concerns which hemisphere the guests occupy. The guests are in predefined regions of the cavity, with the trimethylamine N-oxide **37** sticking to the inwardly directed hydrogen bonds of the pyrroles, while the CHCl$_3$ occupies the original calixarene.[36]

One thorny issue in applications is how to maintain the dynamics of exchange, to stabilize hydrogen-bonded container molecules in polar media without the locked-in forever liabilities of covalent capsules. The pioneering work of the Böhmer[37] group applied the principles of the mechanical bond and produced a locked catenane calixarene capsule. This is optimistically called a "proof of principle"

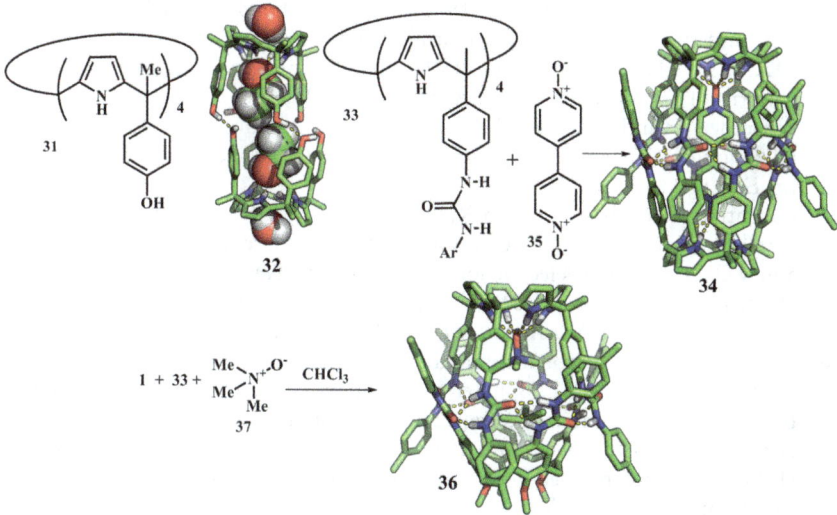

Figure 2.16 *Top*: Calixpyrroles with fixed walls can encapsulate two ethanols in 32 or one complementary large guest as in 35. *Bottom*: A hybrid capsule assembles from a calixarene urea and a calixpyrrole urea. The N-oxide guest is exclusively in the calixpyrrole hemisphere and a $CHCl_3$ solvent is in the resorcinarene hemisphere.

(read: not ready for prime time) but many other catenated capsules were devised.[38] The calixpyrroles have also been catenated.[39]

We are the first to agree that for drug delivery these containers are much more expensive than their potential small molecule payloads (and they have to be small, for no other reason than that the spaces are small). One could imagine applications as a shuttle for a medicine that is loath to cross a biological membrane, but the shuttle must turn over a huge number of times or deliver a catalyst to make them commercially viable. But again, who believed that antibodies would be medicines?

References

1. (a) Gutsche CD. (1989) *Calixarenes*. Royal Society of Chemistry, Cambridge, UK; (b) Vicens J, Böhmer V, eds. (1991) *Calixarenes: A Versatile Class of Macrocyclic Compounds*. Kluwer, Dordrecht; (c) Gutsche CD. (1998) *Calixarenes Revisited*, Royal Society of Chemistry. Cambridge, UK.

2. (a) Etter MC, Adsmond DA. (1990) Hydrogen bond–directed cocrystalliza-
   tion and molecular recognition properties of diarylureas. *J Am Chem Soc*
   112, 8415–8426; (b) Etter MC. (1990) Encoding and decoding hydrogen-bond
   patterns of organic compounds. *Acc Chem Res* **23**, 120–126.
3. Scheerder J, van Duynhoven JPM, Engbersen JFJ, Reinhoudt DN. (1996) Solu-
   bilization of NaX salts in chloroform by bifunctional receptors. *Angew Chem
   Int Ed Engl* **35**, 1090–1093.
4. Casnati A, Pochini A, Ungaro R, *et al.* (1995) Synthesis, complexation, and
   membrane transport studies of 1,3-alternate calix[4]arene-crown-6 conformers:
   a new class of cesium selective ionophores. *J Am Chem Soc* **117**, 2767–2777.
5. Shimizu KD, Rebek Jr J. (1995) Synthesis and assembly of self-complementary
   calix[4]arenes. *Proc Nat Acad Sci USA* **92**, 12403–12407.
6. Böhmer V. (1995) personal communication. Jerusalem.
7. (a) Mogck O, Paulus EF, Böhmer V, Thondorf I, Vogt WJ. (1996) Hydrogen-
   bonded dimers of tetraurea calix[4]arenes: unambiguous proof by single crystal
   X-ray analysis. *Chem Commun* 2533–2534; (b) Mogck O, Böhmer V, Vogt W.
   (1996) Hydrogen-bonded homo- and heterodimers of tetra urea derivatives of
   calix[4]arenes. *Tetrahedron* **52**, 8489–8496.
8. Molokanova O, Podoprygorina G, Bolte M, Böhmer V. (2009) Multiple
   catenanes based on tetraloop derivatives of calix[4]arenes. *Tetrahedron* **65**,
   7220–7233.
9. Hamann BC, Shimizu KD, Rebek, Jr J. (1996) Reversible encapsulation of guest
   molecules in a calixarene dimer. *Angew Chem Int Ed Engl* **35**, 1326–1329.
10. Hamann B. (1990) Ph.D. thesis. MIT Chemistry, p. 170.
11. Castellano R, Rudkevich D, Rebek Jr J. (1996) Tetramethoxy calix[4]arenes
    revisited: conformational control through self-assembly. *J Am Chem Soc* **118**,
    10002–10003.
12. Frish L, Matthews SE, Böhmer V, Cohen Y. (1999) A pulsed gradient spin echo
    NMR study of guest encapsulation by hydrogen-bonded tetraurea calix[4]arene
    dimers. *J Chem Soc Perkin Trans* **2**, 669–671.
13. Cohen Y, Avram L, Frish L. (2005) Diffusion NMR spectroscopy in supramolec-
    ular and combinatorial chemistry: an old parameter — new insights. *Angew
    Chem Int Ed* **44**, 520–554.
14. Schalley CA, Castellano RK, Brody MS, Rudkevich DM, Siuzdak G, Rebek Jr J.
    (1999) Investigating molecular recognition by mass spectrometry: characteriza-
    tion of calixarene-based self-assembling capsule hosts with charged guests. *J Am
    Chem Soc* **121**, 4568–4579.
15. Castellano RK, Rebek Jr J. (1998) Formation of discrete, functional assemblies
    and informational polymers through the hydrogen bonding preferences of cal-
    ixarene aryl and sulfonyl tetraureas. *J Am Chem Soc* **120**, 3657–3663.
16. Castellano RK, Rudkevich DM, and Rebek Jr J. (1997) Polycaps: reversibly
    formed polymeric capsules. *Proc Nat Acad Sci USA* **94**, 7132–7137.
17. Fouquey C, Lehn JM, Levelut AM. (1990) Molecular recognition directed self-
    assembly of supramolecular liquid crystalline polymers from complementary
    chiral components. *Adv Mat* **2**, 254–257.

18. Lomakin A, Benedek GB, Castellano RK, Nuckolls C, Rebek Jr J. (2000) Quasielastic light scattering study of the reversible polymerization of hydrogen-bonded capsules. *Trends Opt Photon* **47**, 27–29.
19. Castellano RK, Nuckolls C, Holger SH, Eichorn MR, Wood A, Lovinger J, Rebek Jr J. (1999) Hierarchy of order in liquid-crystalline polycaps. *Angew Chem Int Ed Engl* **38**, 2603–2606.
20. Castellano RK. (2000) Ph.D. thesis. MIT.
21. Castellano RK, Rebek Jr J. (1998) Formation of discrete, functional assemblies and informational polymers through the hydrogen bonding preferences of calixarene aryl and sulfonyl tetraureas. *J Am Chem Soc* **120**, 3657–3663.
22. Mogck O, Böhmer V, Vogt W. (1996) Hydrogen-bonded homo- and heterodimers of tetra urea derivatives of calix[4]arenes. *Tetrahedron* **52**, 8489–8496.
23. Rebek Jr J. (2000) Host–guest chemistry of calixarene capsules. *Chem Comm* **8**, 637–643.
24. For relevant reviews at that time, see: (a) Selvin PR. (1995) *Methods Enzymol* **246**, 300–334; (b) Yang M, Millar DP. (1997) *Methods Enzymol* **278**, 417–444.
25. Castellano RK, Craig SL, Nuckolls C, Rebek Jr J. (2000) Detection and mechanistic studies of multi-component assembly by fluorescence resonance energy transfer. *J Am Chem Soc* **122**, 7876–7822.
26. Brody MS, Schalley CA, Rudkevich DM, Rebek Jr J. (1999) Synthesis and characterization of a unimolecular capsule. *Angew Chem Int Ed Engl* **38**, 1640–1644.
27. Xu H, Hampe EM, Rudkevich DM. (2003) Applying reversible chemistry of CO_2 to supramolecular polymers. *Chem Commun* **22**, 2828–2829.
28. Rudkevich DM. (2007) Progress in supramolecular chemistry of gases. *Eur J Org Chem* **20**, 3255–3270.
29. Böhmer V, Mogck O, Pons M, Paulus EF. (1999) In M. Pons (ed.). *NMR in Supramolecular Chemistry*. Kluwer, Dordrecht, pp. 45–60.
30. Gonzalez JJ, Ferdani R, Albertini E, *et al.* (2000) Dimeric capsules by the self-assembly of triureidocalix[6]arenes through hydrogen bonds. *Chem Eur J.* **6**, 73–80.
31. Arduini A, Ferdani R, Pochini A, *et al.* (2002) Non-bonded water molecules confined into a self-assembled calix[6]arene cage. *J Supramol Chem* **2**, 85–88.
32. Gale PA, Sessler JL, Kral V, Lynch V. (1996) Calix[4]pyrroles: old yet new anion-binding agents. *J Am Chem Soc* **118**, 5140–5141.
33. Baeyer A. (1872) *Ber Dtsch Chem Ges* **5**, 25.
34. Baeyer A. (1886) *Ber Dtsch Chem Ges* **19**, 2184–2185.
35. Anzenbacher P, Jursikova K, Lynch VM, *et al.* (1999) Calix[4]pyrroles containing deep cavities and fixed walls. Synthesis, structural studies, and anion binding properties of the isomeric products derived from the condensation of *p*-hydroxyacetophenone and pyrrole. *J Am Chem Soc* **121**, 11020–11021.
36. Chas M, Gil-Ramirez G, Ballester P. (2011) Exclusive self-assembly of a polar dimeric capsule between tetraurea calix[4]pyrrole and tetraurea calix[4]arene. *Org Lett* **13**, 3402–3405.

37. Wang LY, Vysotsky MO, Bogdan A, *et al.* (2004) Multiple catenanes derived from calix[4]arenes. *Science* **304**, 1312–1314.
38. Molokanova O, Bogdan A, Vysotsky MO, *et al.* (2007) Calix[4]arene-based bis[2]catenanes: synthesis and chiral resolution *Chem Eur J* **13**, 6157–6170.
39. Chas M, Ballester P. (2012) A dissymmetric molecular capsule with polar interior and two mechanically locked hemispheres. *Chem Sci* **3**, 186–191.

CHAPTER 3

The Cylindrical Capsule

Introduction

We had a number of reasons to explore nonspherical capsule shapes, and a cylindrical shape was particularly appealing. In such a confined space, guest rotational motions could be limited and new arrangements of molecules that were held inside could be envisioned. For the curvature, Thomas Heinz and Dmitry Rudkevich chose the accessible and abundant resorcinarene developed by Högberg,[1] as will be described later. This module has assembly properties as a hexameric capsule which warrants a chapter all of its own, but at the time this feature had not been recognized. What was known was that the resorcinarene octol could react smoothly with activated aromatic dihalides — a reaction developed by Cram[2] and Dalcanale[3] — to give a deep vase-like cavitand **3.1** with the four aromatic walls defining the space (Figure 3.1).

This is not the only shape, however. This vase-like structure is in equilibrium with a kite-like form **3.2** with the four walls flipped outward, which is the more stable isomer in most organic solvents. To make matters of curvature worse, the kite is unaccountably self-complementary in shape: two molecules come together, squeeze the solvent out from between them and stick together through aromatic stacking attractions to form "velcrands" **3.3**.[4] Although occasional molecular recognition studies[5] were reported with the deep cavitands,[6] their tendency to dimerize limited their use.

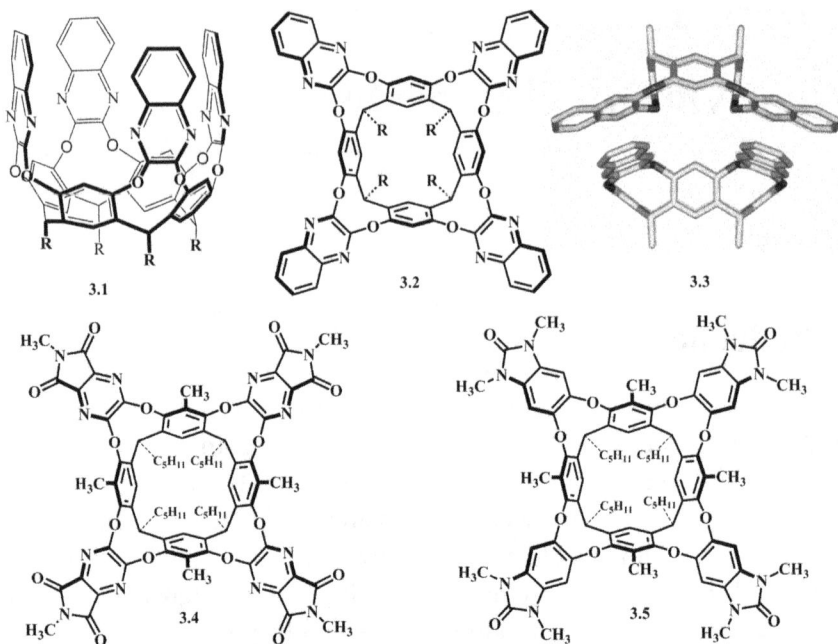

Figure 3.1   *Top*: Vase **3.1** and kite **3.2** forms of a deep cavitand and the dimeric kite velcrand **3.3**. *Bottom*: Tetraimide **3.4** and tetraurea **3.5** kite structures; the N-methyl groups prevent aggregation through hydrogen bonding.

Capsules

Two particular structures studied as velcrands by Cram were the tetraimide **3.4** and the tetraurea (benzimidazolone) **3.5**, which bore methyl groups on all the N-atoms, as well as C-methyl groups on the resorcinarene. The C-methyls were believed to give a lock-and-key fit in the dimeric velcrands but, as we have described elsewhere,[7] we felt that the N-methyl groups were unnecessarily preventing new assemblies. We will describe the modifications of **3.5** later, but the removal of eight methyls from **3.4** by Heinz and Rudkevich exposed a structure which recoalesced as a cylindrical capsule! The tetraimide cavitand **3.6** features a self-complementary hydrogen bond array along its upper rim (Figure 3.2).[8] When two of these are brought together in their vase-like conformations and concave face to concave face, the capsule is formed. It is held together by eight bifurcated hydrogen bonds and it readily assembles in organic solvents. The figure shows

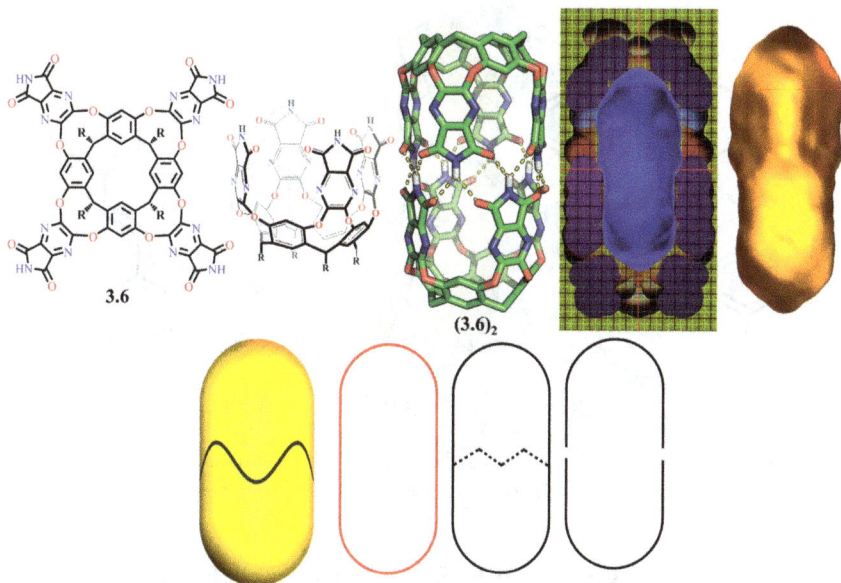

Figure 3.2   *Top*: Depictions of the cavitand **3.6** and various renderings of its dimeric capsule (**3.6**)$_2$ discussed in this chapter. The blue area in the cross-section figure represents the shape of the space inside; it is defined by two square prisms rotated 45° with respect to each other, as emphasized in the golden figure. *Bottom*: Various depictions of the capsule used elsewhere in this chapter.

a space-filling and a cross-section view of the capsule. The volume is about 425 Å^3, but the *shape* of the space lends this capsule its desirable properties. The space is two square prisms rotated 45° from each other; the tapered ends are provided by the space of the resorcinarenes, also rotated 45° with respect to the square prisms. The peripheral "feet" of the resorcinarenes are typically n-C$_{11}$H$_{23}$ and were chosen for ease of synthesis and solubility in organic solvents. These are deleted in the figures and structures that follow.

Single Large Guests

Rigid Guests

The capsule binds molecules with congruent sizes, shapes and compatible chemical surfaces. For rigid guest molecules, it shows an

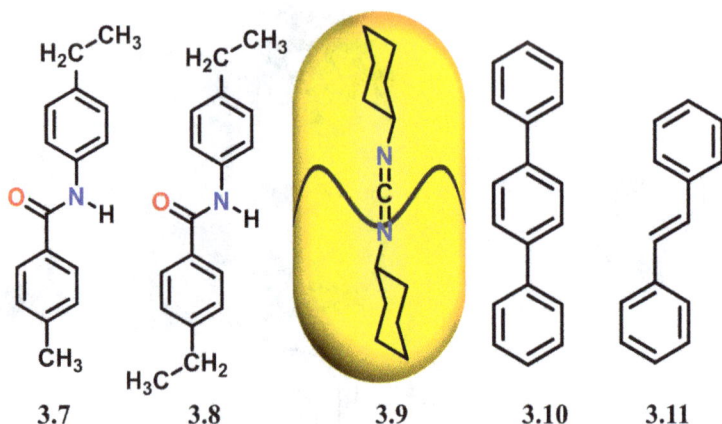

Figure 3.3 Typical rigid molecules used as "rulers" of the capsule's capacity. Anilide **3.7** is accommodated in the capsule, but the slightly longer **3.8** is not.

exquisite selectivity for length. Steffi Körner and Fabio Tucci used benzanilides as molecular rulers and found that with a methyl at one *p*-position and an ethyl at the other (Figure 3.3, structure **3.7**), the molecule fits inside. But a benzanilide with two *p*-ethyl groups (**3.8**) no longer fits; no assembly took place.[9] Moreover, substitution at the N-atom of the anilide showed that N-methyl and N-ethyl groups could be tolerated, but N-propyl or larger groups were not encapsulated.

Molecules such as dicyclohexylcarbodiimide **3.9** were very good guests, and we will see (Chapter 8) that its reactivity is turned off while in the capsule, but that unusual chain reaction kinetics can be observed when appropriate reagents outside the capsule are present. Molecules like terphenyl **3.10** and *trans*-stilbene **3.11** are bound, but *cis*-stilbene is not.[10]

Flexible Guests

Amino acids

One of our first encounters with the effects of encapsulation on guest shape was with protected amino acid and peptide derivatives. Osamu Hayashida and Lubo Sebo examined[11] increasingly long esters of

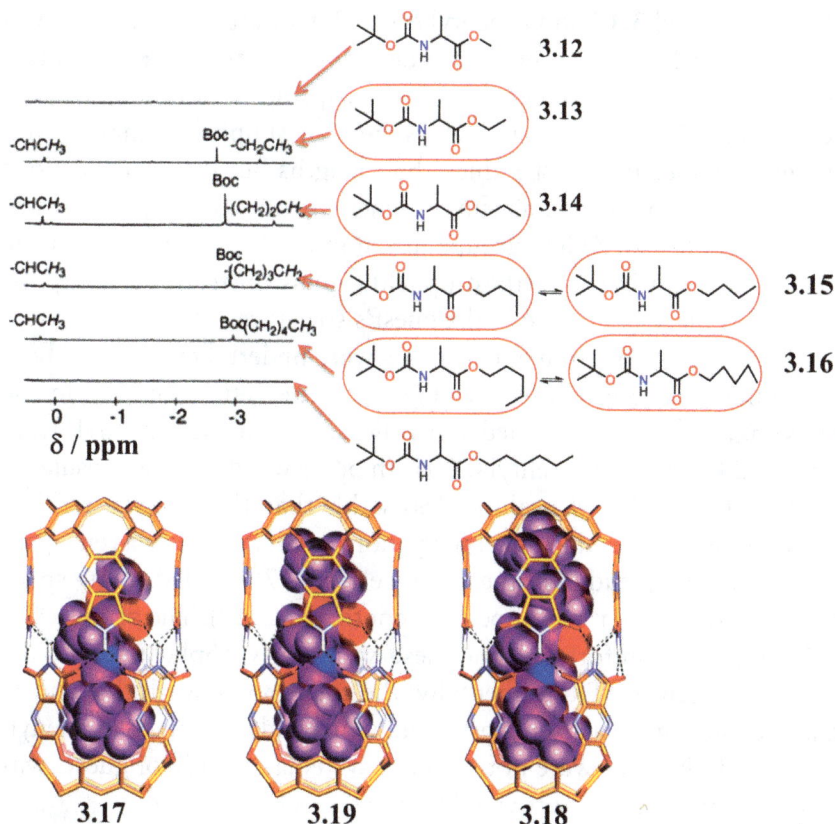

Figure 3.4 *Top*: Boc-L-alanine esters as guests. The complex with the longest guest, the *n*-pentyl ester **3.16**, is 4 kcal/mole less stable than the propyl ester **3.14**. *Bottom*: Modeling of the methyl ester **3.17** shows empty space inside the capsule while the hexyl ester **3.18** is folded. The best fit is the propyl ester **3.19**.

Boc-L-alanine (Figure 3.4). Among this series the shortest (methyl ester **3.12**), with a molecular length of 10.9 Å, was not encapsulated but the slightly longer guests (ethyl **3.13** and propyl **3.14**) were. The encapsulated Boc and terminal alkyl groups showed large upfield shifts ($\Delta\delta$) of -4.15 to -4.30 and -4.40 to -4.43 ppm, respectively, which place them at the ends of the capsule, while the methyls of the amino acid side chain showed smaller upfield shifts ($\Delta\delta$ -0.94 to -0.97 ppm), which locate them in the middle of the capsule, near the seam of hydrogen bonds.

The *n*-butyl **3.15** and *n*-pentyl esters **3.16** were also taken in even though their lengths in their extended conformations are equivalent to or even exceed the dimensions of the capsule. The terminal methyl resonances of the esters were always the highest upfield signals. To fit, the guest molecules must reduce their lengths but their contractions are not through curling, as shown schematically in Figure 3.4. Such curling would place CH_2 groups in the most shielded regions of the hosts and result in the furthest upfield signals. Instead, the compression takes place elsewhere in the guest's structure.

Pairwise competition experiments were undertaken with the Boc-L-alanine *n*-alkyl esters to determine the energetics involved in the buckling, and these revealed the following sequence: propyl 1.00: butyl 0.24: ethyl 0.03: pentyl 0.001. In other words, the encapsulated pentyl ester **3.17** is ~4 kcal/mol less stable than the propyl ester. For the alanines, neither the longer hexyl nor the benzyl ester was encapsulated. Modeling studies of the methyl ester **3.17** revealed empty space while the hexyl ester was hopelessly contorted (**3.18**). The propyl ester **3.19** appears comfortable but these were early applications of the available software. Experimentally, the corresponding glycine and β-alanine esters were also encapsulated, while wider (valine and phenylalanine) derivatives were not. A more systematic study of guests was desirable and this was provided by the series of *n*-alkanes.

Alkanes

The size of the cavity is a matter of some uncertainty and depends to some extent on software that is used to calculate it. Typically, a probe is rolled over the inner surface and, with that, the dimensions are given as about 16.5 Å inside and slightly more than 7 Å in width. The encapsulation of *n*-tetradecane occurs spontaneously from mestylene-d_{12} and gives a spectacular first-order NMR spectrum, where signals of each methylene and methyl group are widely separated.[12]

This separation of signals is a consequence of the 16 aromatic panels of the capsule; they create a magnetic anisotropy that shield nuclei inside from applied magnetic fields outside. The result is the upfield shift of NMR signals for encapsulated guests and the shifts

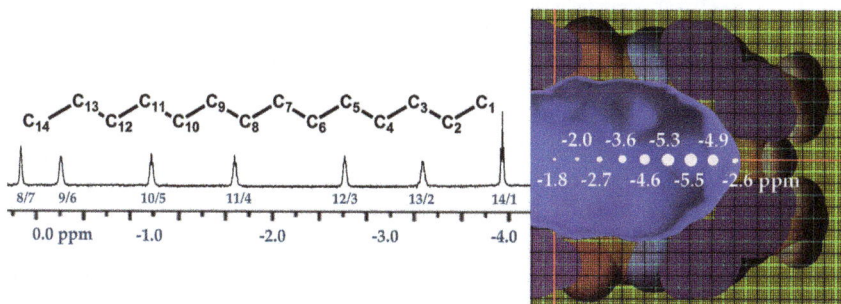

Figure 3.5 *Left*: The NMR spectrum of encapsulated C14. *Right*: Calculated nucleus independent chemical shift (NICS) values along the central axis of the capsule. [Reprinted with permission from *J Am Chem Soc* **126**, 13512–13518. Copyright 2004, American Chemical Society.]

are dramatic: as seen for C14, the methyl signal appears at −4 ppm or an upfield shift (Δδ) of nearly −5 ppm from outside to inside. These large shifts place encapsulated alkane ¹H NMR signals in a part of the spectrum far upfield from the TMS standard and in a region clear of other signals — a favorable circumstance.

It is possible to calculate these chemical shifts using the predictions of the nucleus-independent chemical shifts (NICS) developed by Schleyer.[13] The calculated NICS results are shown in Figure 3.5 for the capsule along its central axis, but it should be understood that the effects grow larger as the nuclei move closer to the walls. Even so, the trend is clear: the maximum shift (−5.5 ppm) occurs in the resorcinarene cavitand and the effect decreases as the nuclei move toward the center of the capsule or to its very tapered end. The 5 ppm shift observed for the methyl groups of C14 place them near the rim of the cavitand.

Regardless of the software used, *n*-tetradecane is too long a molecule to fit inside in its fully extended form. Instead, modeling shows that it can coil into a helix to be accommodated inside the structure (Figure 3.6). This coiling involves a number of *gauche* interactions along the chain, each one of which raises the energy by 0.55 kcal/mol in the liquid state.[14] However, this price is apparently paid by the hydrogen bonds of the capsule complex and attractive C–H/π interactions between guest and host. The coiled molecule is

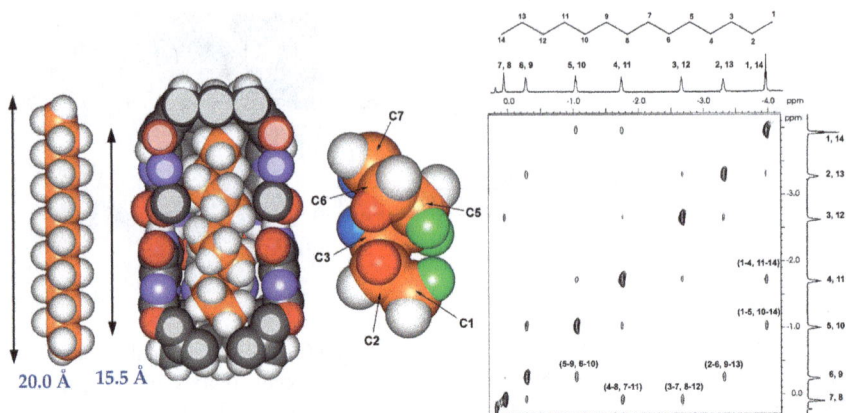

Figure 3.6 *Left*: Dimensions of extended C14 and its helically coiled conformation in the capsule. Coiling brings H-atoms on C_1 close to those on C_5 (green); those on C_2 close to C_6 (red); etc. *Right*: The 2D NOESY NMR spectrum at 600 MHz and 300 millisecond mixing time, showing cross-peaks between H-atoms on C_1 and C_5, C_2 and C_6, etc. [Reprinted with permission from *J Am Chem Soc* **126**, 13512–13518. Copyright 2004, American Chemical Society.]

shorter and thicker, and it was simply fortunate that the dimensions of a helically coiled alkane nicely match the cavity dimensions. The *gauche* interactions were detected by Alessandro Scarso and Laurent Trembleau by using 2D NOESY NMR experiments.[15] The spectrum showed cross-peaks between H-atoms on C_1 and C_5, C_2 and C_6 and so on up the chain, as expected if C14 is coiled inside the capsule.

The coiling of C14 is not without an energetic price. In competition experiments with a series of shorter and less compressed alkanes, Wei Jiang determined that the price of coiling — between the fully extended C11 and the tightly coiled C14 — was ~6 kcal/mol.[16] However, this figure includes the improved C–H/π attractions and the internal pressure that the C14 exerts on the ends of the capsule (Table 3.1).

The encapsulated C14 is not fixed in a chiral helix; instead, it is dynamic and interconverting between the two enantiomeric forms very rapidly on the NMR timescale in a motion that we have been unable to freeze out. There is a reasonable mechanism where it could do so with only a small change of length. The *trans*-alkene,

Table 3.1 *Gauche* Interactions, Relative Binding Affinities, Relative Free Energies and Packing Coefficients of *n*-Alkanes in the Capsule.

Guest	Gauche	K_{rel}	$\Delta\Delta G^0$ (300 K, kJ mol^{-1})	PC
n-nonane	—	0.3	3.0	43
n-decane	—	16.9	−7.1	48
n-undecane	0	100	−11.5	52
n-dodecane	4	24.4	−8.0	54
n-tridecane	8	1.0	0.0	55
n-tetradecane	11	0.008	12.0	58

7-tetradecene is also accommodated with an affinity comparable to that of C14. This suggests that a crankshaft-like motion along the alkane is possible inside the capsule: untwisting from (+) *gauche* to antiperiplanar (similar to a *trans* double bond) and then twisting to (−) *gauche* could take place anywhere along the chain. The change in length from a *gauche* to an extended alkyl chain is only 0.32 Å.[17] When this rotation is repeated along the C14 chain, the sequence interconverts the enantiomeric helices. A chiral helix has been observed in this capsule by Waldvogel,[18] who used 2-tetradecanol as a guest to induce local, stereoselective helical folding. A theoretical study also predicts helical coiling of alkanes in carbon nanotubes of appropriately narrow bore[19] (Figure 3.7).

Complex formation takes place on mixing the components (within seconds). A competition between six alkanes (C9–C14) showed that the equilibration occurred in a sequence (Figure 3.7), with an initial uptake of C9, and then its displacement by C10, which is slowly replaced by the most favored C11. The rates of uptake are clearly length-dependent. We will deal with exchange rates of other guests later in this chapter.

Alkynes

Addition of one methylene to the guest gives *n*-pentadecane (C15), but no encapsulation occurred: there is nothing that C15 can do to make itself fit inside this capsule. However, the methyl groups

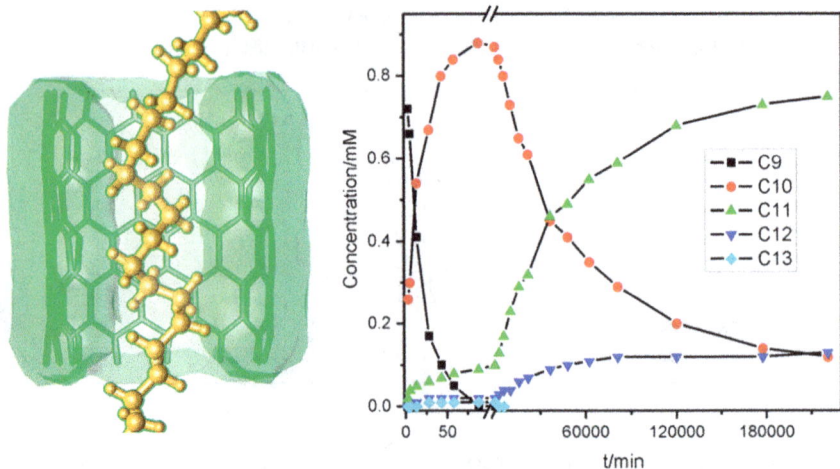

Figure 3.7 *Left*: An alkane modeled inside a narrow bore carbon nanotube. [Reproduced in part from *Phys Chem Chem Phys* (2014), **14**, 2707–2709, with the permission of The Royal Society of Chemistry]. *Right*: Distribution of encapsulated alkanes with time during a competition experiment with C9–C13. The shorter alkanes are taken up faster but are eventually replaced by C11. [Reprinted with permission from *J Am Chem Soc* **134**, 8070–8073. Copyright 2012, American Chemical Society.]

of pentadecane are relatively "blunt" and do not fit well into the tapered ends of the resorcinarenes. Tetsuo Iwasawa and Dariush Ajami decided to make one end of the alkane narrower — sharpening it to a point — such as in a primary acetylene. They found that, indeed, encapsulation of 1-pentadecane occurred[20] and resulted in a clear first-order NMR spectrum (Figure 3.8). The needle-like primary acetylene C–H tip was able to access the tapered ends, and penetrate places that excluded a methyl group.

We examined a series of primary acetylenes separated by a biphenyl spacer from increasingly large alkyl groups. The larger groups forced the primary acetylene deeper into the cavitand toward the narrower end. The series showed the NMR signal for the acetylenic hydrogen moving *downfield* as it approached the narrow opening at the end of the capsule. This is in complete accord with Schleyer's NICS calculations. Calculated structures that correspond to the spectra are shown in Figure 3.9.

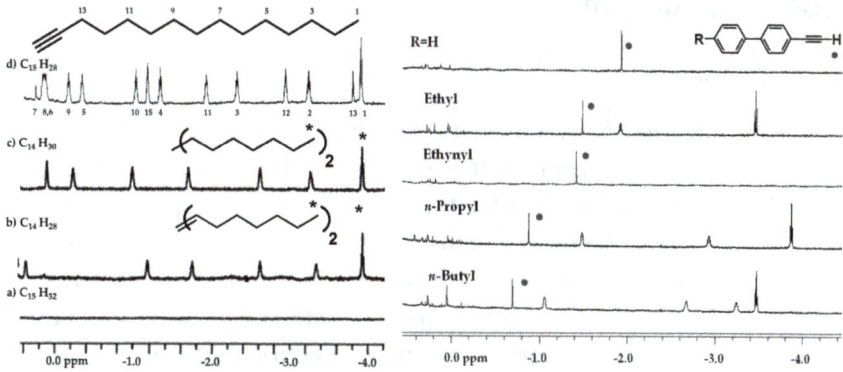

Figure 3.8 *Left*: NMR spectra of encapsulated hydrocarbons. Both C14 and its 7-*trans*-alkene are bound; C15 is not accommodated but its primary acetylene fits inside. *Right*: NMR spectra of increasingly long rigid alkynes. The signal of the acetylenic H-atom moves downfield as it approaches the end of the space. [From *Proc Natl Acad Sci USA* **103**, 8934–8936. Copyright 2003, National Academy of Sciences of the United States of America.]

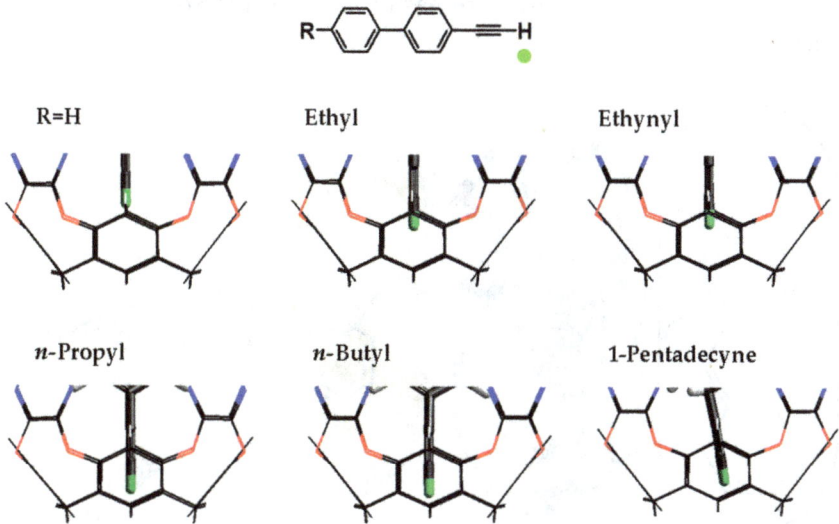

Figure 3.9 Calculated structures of encapsulated alkynes, showing the approach of the acetylenic H-atom to the ends of the capsule. [From *Proc Natl Acad Sci USA* **103**, 8934–8936. Copyright 2003, National Academy of Sciences of the United States of America.]

Coencapsulation

Social Isomers

However, the most intriguing aspect of the capsule that results in some of its unique properties is that it has the ability to select and encapsulate two *different* molecules that together fill the space properly. For example, in a mixture of benzene and *p*-xylene, a single capsule is formed that has one molecule of each inside. We will use this in subsequent chapters to generate new forms of stereochemistry and chiral spaces, but for now we refer to this metaselection as coencapsulation. The coencapsulated molecules can be very different. For example, Alex Shivanyuk and Alessandro Scarso determined that anthracene alone does not go into the capsule: one molecule of anthracene does not fill enough space, and two molecules fill too much. However, when a solution of anthracene and the capsule is exposed to the slow stream of methane, a single capsule complex **3.20** (Figure 3.10) emerges with one molecule of the gas and one molecule of the anthracene inside.[21] Many small molecule gases are easily coencapsulated with a larger guest.

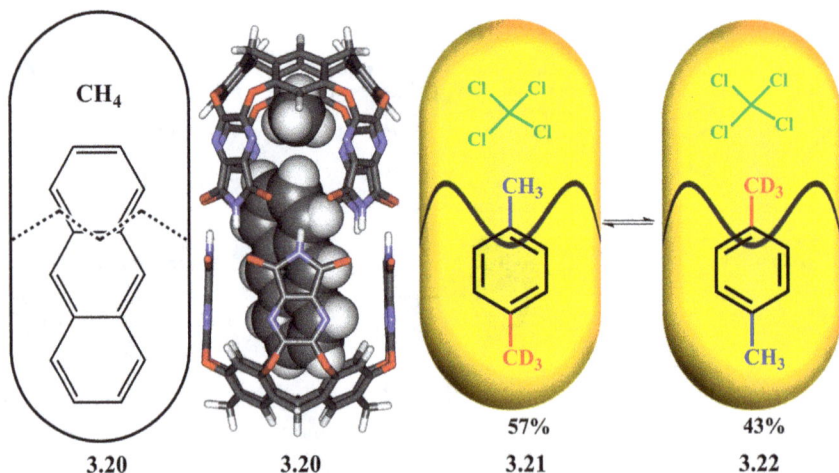

Figure 3.10 *Left*: Renderings of coencapsulated methane and anthracene; neither guest is encapsulated as a single entity. *Right*: Isotopomers of partially deuterated *p*-xylene, showing the preference of the CD_3 group for the resorcinarene.

It should be noted that the most useful solvent for NMR studies of this capsule is deuterated mesitylene, since this molecule does not fit well inside and is the largest commercial NMR solvent available. Its expense and impurities are disadvantages — even a 0.1% contaminant is at millimolar concentrations. Screening for possible impurities in the commercial solvent revealed that the resting state of the capsule contained one molecule of deuterated *p*-xylene coencapsulated with one molecule of deuterated benzene. Fortunately, these could be replaced by intended guests at higher concentrations.

Isotopomers

Many coencapsulated combinations have been explored, and one of them allowed the measurement of an isotope effect inside the small confined space. Alessandro Scarso and Dalit Rechavi coencapsulated carbon tetrachloride with *p*-xylene, but the *p*-xylene had one of the methyl groups deuterated.[22] The two possible arrangements are social isomers (Chapter 5) but were not there in equal amounts (Figure 3.10). In fact, the deuterated methyl group preferred to make contact with the resorcinarene at the tapered end of the cavity: isotopomer **3.21** is more stable than **3.22**. Computations by our collaborator Ken Houk showed that such a CD_3 group has more attractions to the aromatics than the simple CH_3 group.[23]

The isotope effects observed in the encapsulation phenomenon remain something of a mystery. The data are limited to those cases where guests' CH bonds are replaced by CD bonds. Our experience and that of Haino are that deuterated versus protiated molecules prefer to be partitioned into capsules from organic solvents.

Haino found that the relative binding affinities (K_D/K_H) of encapsulated guests in $CDCl_3$ were inevitably larger than unity, and so the supramolecular complex with the deuterated guest is thermodynamically more stable than with the protiated one.[24] Warmuth described chromatographic differences of protiated versus deuterated guests in covalent carceplexes.[25] Gibb deuterated internal sites on a host cavitand and found that it bound halogenated guests better in DMSO-d_6 than the protiated cavitand.[26] However, the experience appears to be different in aqueous media. Bergman and Raymond have studied the encapsulation and the exchange of carbon hydrogen versus carbon-deuterated molecules in water (D$_2$O) and have come to the opposite conclusion. That is to say, C–H molecules are preferentially encapsulated versus C–D molecules, for both interior and exterior binding of a benzylphosphonium guest in aqueous solution.[27] The preferred association of protiated guests is driven by enthalpy and opposed by entropy.

Anions

In a survey of different types of molecules encapsulated here, Osamu Hayashida and Alex Shivanyuk found that certain anions were welcome guests, if the solution outside was chloroform (Figure 3.11).[28] Ordinarily, three molecules of chloroform occupy the capsule, but anions such as tosylate **3.23**, BF$_4^-$ **3.24**, PF$_6^-$ **3.25** and halides are all taken up from the solvent into the capsule and are coencapsulated with the remaining chloroform molecules.

Apparently, the anions are not happy in that solvent and prefer to be inside the capsule near the polar seam of hydrogen bonds. Surprisingly, in a solution of tetraethyl ammonium hexafluorophosphate,

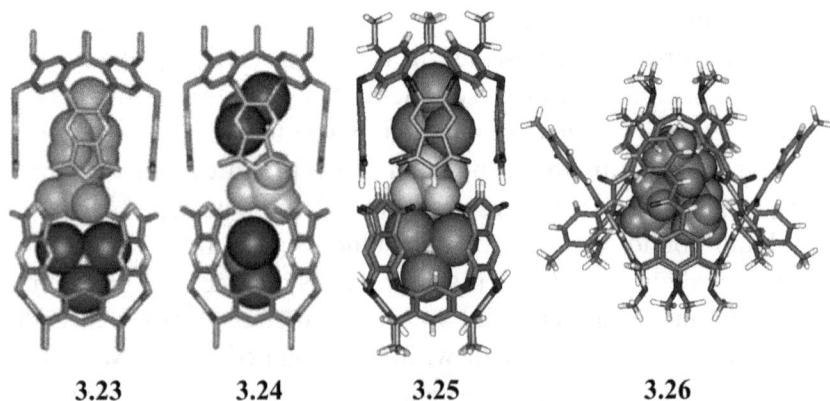

| 3.23 | 3.24 | 3.25 | 3.26 |

Figure 3.11   Anions like tosylate and fluoborate are encapsulated with chloroform solvent molecules. Hexafluorophosphate and tetraethyl ammonium ions can be simultaneously encapsulated in separate containers.

the cylindrical capsule and the calixarene capsule showed no free ions in the NMR spectrum. Both ions were encapsulated in their cavities even though a distance of some 10 Å separates these charges in **3.25** and **3.26**.

Hindered Rotation

One of the largest molecules we found would fit was 2,2 paracyclophane **3.27** and a model of its complex **3.28**, shown in Figure 3.12. It is easily seen to be a visually satisfying square peg in a square hole. Now, the guest molecule already has lost its translational freedom inside the capsule, and would resist losing rotational freedom as well. But if rotation is to occur, something has to give.

Alessandro Scarso and Hideki Onagi showed that rotation inside does occur, through dynamic NMR experiments, with the barrier to rotation at ~15 kcal/mol.[29] This was the only instance we have encountered of hindered rotation on this axis of the capsule. That the rotation occurs at all is due to a "breathing" motion of the center of the capsule by distortion of the hydrogen bond seam. The paracyclophane spins along the long axis of the capsule and the widening

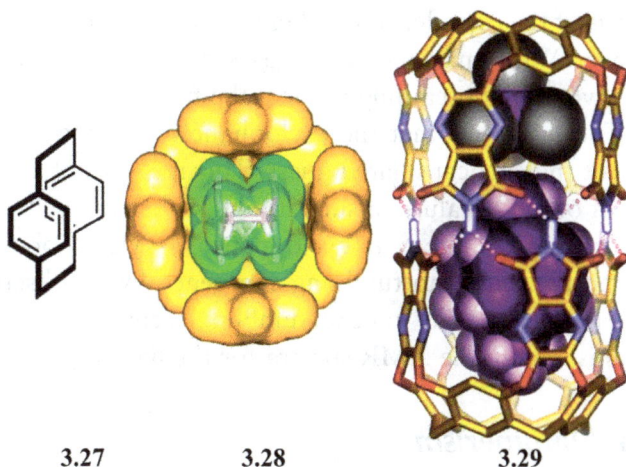

3.27   3.28   3.29

Figure 3.12   Structure of the rigid paracyclophane **3.27** and its resting state in one end of the capsule **3.23**. It can be coencapsulated with a large guest such as CCl_4, as shown in **3.29**.

Table 3.2 Effects of the Co-guest and Packing Coefficient (PC) on the Rotation Rate (Temperature = 263 K) and Activation Energy $\Delta G^{\ddagger}$ of *p*-cyclophane in the capsule **1•1**. [Reprinted with permission from *J Am Chem Soc* 126, 12728–12729. Copyright 2004, American Chemical Society.]

Co-guest	Volume (Å^3)	PC (%)	k (min^{-1})	$\Delta G^{\ddagger}$ (kcal/mol)
CH_3CH_3	42	62	178	14.8
$(CH_2)_3$	52	64	257	14.6
$(CH_3)_2CO$	60	66	297	14.5
$CHCl_3$	75	69	127	15.0
$(CH_3)_2CHCl$	76	70	76	15.2
$(CH_3)_2CHBr$	84	72	75	15.2
CCl_4	91	73	65	15.3
$CHBr_3$	99	75	36	15.6
C_6H_{12}	97	75	23	15.9
CCl_3Br	114	79	46	15.5

of the capsule reduces the "friction" involved. Unexpectedly, the rate of spinning depends on the coencapsulated guest (Table 3.2). With a large guest such as CCl_4 **3.29**, the cyclophane is forced toward the tapered ends of the capsule, where there is more friction and the spinning is slower. With small guests like gases, the cyclophane floats near the center, where the breathing reduces the friction and the spinning is faster. The upshot is that the cyclophane's spinning rate reports the "size" of the coencapsulated guest. This measure of size is quite independent of the *A* values of substituted cyclohexanes.[30]

With molecules like paracyclophane, *p*-xylene or other *p*-disubstituted benzenes, no tumbling can generally be observed, but with shorter molecules like toluene, tumbling occurs and only an averaged signal is seen in the NMR spectra for the methyl group.

Keto–Enol Tatomerism

In 2004 my colleague Albert Eschenmoser and I were discussing the ketone–enol equilibrium and he quoted me a statement that was something like "the choleic acid complexes with ketones prefer the enolic form". This point is brought up by the Fiesers[31] in their textbook on steroids, and (to my knowledge) is also the first description of encapsulation since "a guest which is not covalently bonded to a host" is mentioned. Bile acids like choleic acid were known to dissolve certain hydrophobic molecules and were believed to do so as stoichiometric complexes. We examined choleic acid complexes by NMR, but could not find any discrete structures; only ill-defined aggregates were present in solution. Earlier, Fujita[32] had encapsulated a β-diketone, exclusively in the enol form; we were interested to see if this was a general phenomenon that extended to our capsules. It was well documented that solvation played a role for β-ketoesters: polar protic solvents that compete strongly for the donor and acceptor sites disrupt the intramolecular hydrogen bond of the enol form and shift the equilibrium toward the keto isomer.[33] This raised the question of how the isolation provided by the mechanical barriers of encapsulation would affect the equilibrium, and Alessandro Scarso took up the appropriate experiments.

We examined a number of keto–enol tautomers inside the capsule, both alone and as a function of various coguests. The equilibrium constant for the single guest benzyl acetoacetate $K_{outside}$ [enol]/[ketone] $= 0.4$ in the solvent outside the capsule but 3.6 inside the capsule. In this complex the α-carbon is surrounded by aromatic surfaces and favors the enol form compared to the free guest in solution. But the nine-fold increase in the equilibrium constant due to encapsulation appears too large to arise only from solvation effects, particularly since the outside solvent (mesitylene d_{12}) is also apolar.

There are numerous equilibria involved for two guests and the difficulties of analysis are exacerbated by social isomerism, as shown in Figure 3.13. Many of these constants were determined. Correlations could be made between the equilibrium constant and the co-guest size, where larger guests favored the ketone. But since the capsule acts as an environment with a gradient of polarity, no profound or even predictable effects were seen.

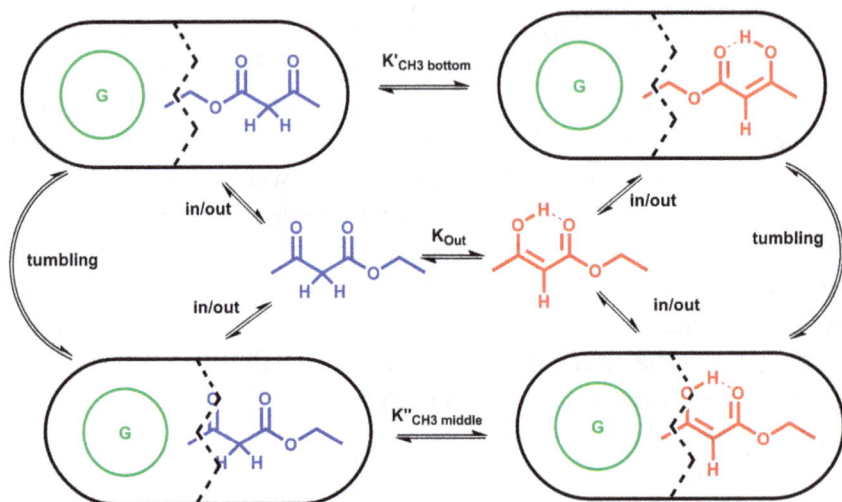

Figure 3.13 Keto–enol isomerization as a function of coencapsulated guests. Several simultaneous equilibria complicate the analysis.

Guest–Guest Attractions

The interactions of co-guests are most directly observed through the effects of hydrogen-bonding and halogen-bonding. These are weak intermolecular forces but are moderately directional, and the cylindrical capsule appeared to be an ideal space for donors and acceptors to be matched: the shape of the space allows the alignment of functional groups and the small space of the capsule results in higher concentrations of both components. Accordingly, both the molarity and the effective molarity are enhanced inside the capsule.

Hydrogen bonds

It was possible to stabilize hydrogen-bonded complexes within the capsule. Molecules such as carboxylic acids are extensively hydrogen-bonded as dimers, chains and oligomeric aggregates in noncompeting solvents. The rapid exchange of partners in solution prevents the observation of discrete species even with fast spectroscopic methods. Benzoic acid binds nicely inside this capsule, as its discrete hydrogen-bonded dimeric structure. The *p*-toluic acid dimer is too

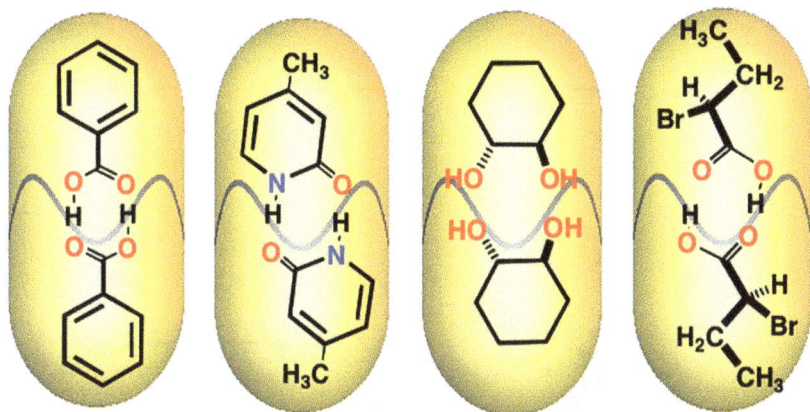

Figure 3.14 *Left*: Hydrogen-bonding dimeric complexes inside the capsule. *Right*: The capsule prefers mirror image isomers of the diol, but identical molecules of the bromoacid.

long to fit inside this capsule and is excluded. Likewise, 2-pyridone is encapsulated as its dimer (Figure 3.14). A number of closely related molecules can be isolated and observed, including mirror image (the cyclohexane diols) and identical (2-Br-butyric acids) dimers; these will be discussed in the chapter on stereochemistry.

Both benzoic acid and its primary amide are bound in the capsule as discrete dimers, and a mixture of the guests gave a third species — the heterodimeric complex (Figure 3.15). This allowed Wei Jiang and Konrad Tiefenbacher to perform an experiment that is not possible in solution, where rapid exchange of all partners occurs and gives only averaged signals in the NMR spectra. The experiment is the disproportionation equilibrium of primary amide (A.A) and carboxylic acid (C.C) homodimers to the heterodimers (A.C) in the capsules. A purely statistical value for the dimensionless constant, $K_d = [A.C]^2 / ([A.A] \times [C.C])$, is 4. Pairing the stronger donor C with the stronger acceptor A would be expected, but what are the hydrogen-bonding preferences inside the capsule?

The carboxylic acid homodimer is preferred in the capsule to its amide counterpart and $K_d = 0.4$.[34] In solution, primary amide homodimers can form hydrogen-bonding networks with each other, through the N–H donor which is not involved in the cyclic array.

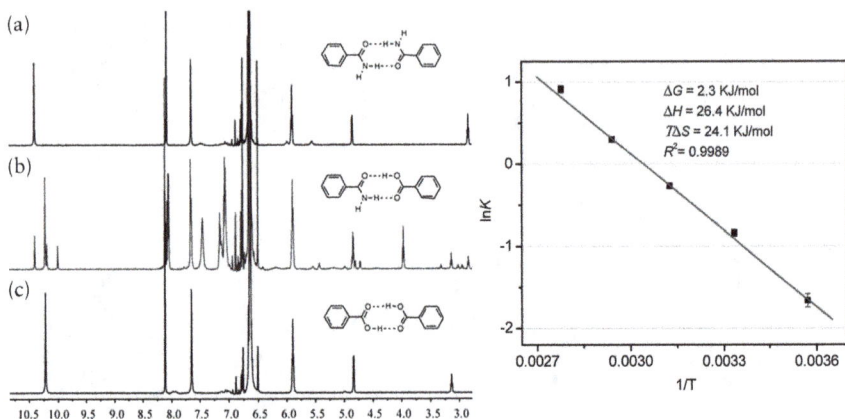

Figure 3.15 The disproportionation of symmetrical dimers can be quantified inside the capsule. In solution, rapid exchange of partners thwarts the analysis. [Reproduced from *Chem Sci* 3, 3022–3025, with permission from The Royal Society of Chemistry.]

Figure 3.16 Geometric details of the acid–amide heterodimeric structure and its shape. [Reproduced from *Chem Sci* 3, 3022–3025, with permission from The Royal Society of Chemistry.]

Inside the capsule there are no acceptors positioned to pair with this exocylic donor. The isolated amide homodimer and its heterodimer are unhappy inside. The van 't Hoff plot shows that the disproportionation is both enthalpically and entropically disfavored.

The asymmetry of hydrogen-bonding in the heterodimer may also contribute to this behavior. The heterodimer (Figure 3.16) is slightly "bent" (173°) with the unequal hydrogen bond length (N–H$\cdots$O, 2.8 Å; O–H$\cdots$O, 2.6 Å). The homodimers have D_{2h} symmetry and their structures are "linear." This small difference could cause steric clashes of the rigid phenyl groups with the capsule walls which weaken the hydrogen bonding both of the capsule and in the heterodimeric guest.

A large change in preferences occurs with the corresponding cyclohexyl amide and acid, which shows that $K_d = 18$. The cyclohexyl guests in the same capsule show a very favorable enthalpy, suggesting that the best donor and acceptor are paired up in the heterodimer. The more flexible cyclohexyl groups can adapt to the shape differences between heterodimers and homodimers and do not apply stress to the capsules. This is consistent with the downfield shift of the capsule's hydrogen bonds. Again, subtle differences in geometry are amplified by the small spaces in the capsules. We will return to this subject in the chapter on expanded capsules.

Halogen bonds

Halogen bonds, like hydrogen bonds, have a moderate directionality as they involve a donor, usually a Br or I atom presenting a "sigma hole" to a Lewis base acceptor (Figure 3.17).[35] The repulsion between the lone pairs of the halide and the Lewis base enforce the directionality,[36] and the resulting rectilinear geometry is ideal for alignment in the cylindrical capsule. Golam Sarwar and Dariush Ajami modeled appropriate halogen-bonding pairs for encapsulation.

As halogen bond donors, they used perfluoroalkyl iodides; and as acceptors, pyridine derivatives. Two molecules of perfluoropropyl iodide are coencapsulated and the chemical shifts indicate that iodides are directed toward the ends of the capsule (Figure 3.18). There is halogen-bonding between the aromatic faces[37] of the resorcinarene

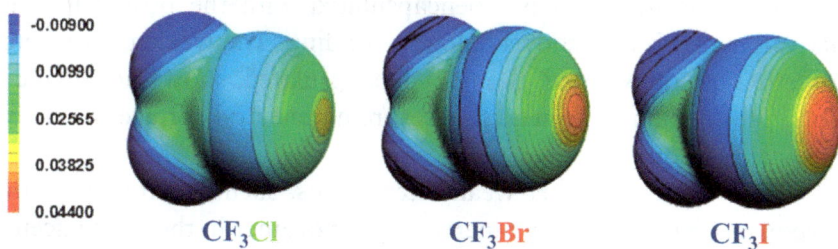

| -0.00900 |
| 0.00990 |
| 0.02565 |
| 0.03825 |
| 0.04400 |

CF_3Cl CF_3Br CF_3I

Figure 3.17   Electrostatic surfaces of halo carbons, with the sigma hole in red.

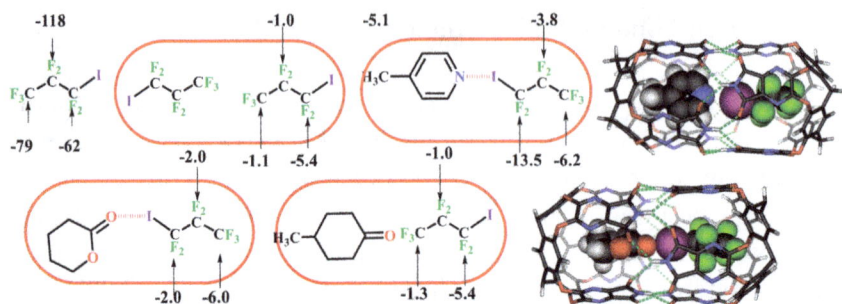

Figure 3.18   Chemical shifts of ^{19}F NMR spectra for iodo alkanes. The changes on encapsulation ($\Delta\delta$) are given in ppm.

and the sigma hole of the iodide. However, when γ-picoline is added, one of the iodides is ejected and the other flips in the capsule's space in order to make a halogen bond with the lone pair of the nitrogen. This is accompanied by very large upfield shifts in the ^{19}F NMR spectra (Figure 3.18). The numbers on the capsule cartoons are the observed $\Delta\delta$ in the NMR spectra and represent the combined results of halogen binding and the NICS calculations. The orientation is very important in this case; neither β- nor α-picoline shows any binding to the iodide inside the capsule because they cannot present their N to the halide along the axis of the capsule. The length can also be determining in the space: a longer donor requires a shorter acceptor. Specifically, perfluorobutyl iodide is only coencapsulated with pyridine itself, and again shows a very strong binding inside the capsule. The computations of our collaborators Giannoula Theodorakopoulos and Ioannis Petsalakis gave the figures shown.[38]

Even a lactone can be coencapsulated with the propyl iodide derivative and it also shows halogen-bonding. The weaker acceptor, 4-methyl cyclohexanone, though coencapsulated, shows no interaction with the halide. Instead, the resorcinarene aryl surface provides the halogen bond acceptor.

There are many other weak interactions, such as halogens with carbonyls, that are showing increasing importance in the pharmaceutical industry. These forces could likewise be enhanced and amplified in the small spaces of the capsule. Again, the shape of the space allows the alignment of functional groups and holds them in place for long times. These features are lost in bulk solution due to random collision encounters and rapid exchange of partners through diffusion.

Guest Exchange

Small Molecules

How molecules get in and out of this capsule was initially studied by Stephen Craig and Shirley Lin.[39] Not unexpectedly, their experiments revealed a mechanistic continuum for exchange. The structure of the guest determines the sequence and rate-determining steps but does not involve the creation of empty volumes or complete dissociation of the capsule. Initially the capsule with benzene and coencapsulated *p*-xylene was examined. The NMR EXSY experiments performed at different benzene concentrations revealed a process with concentration dependence, but the nonzero intercept in Figure 3.19 speaks for a second, concentration-independent mechanism. Surprisingly, the two halves of the capsule do not exchange environments during the benzene exchange: the incoming benzene resides in the same half of the capsule vacated by the outgoing benzene! The xylene does not move during the benzene exchange processes. We propose that flaps in the capsule open to allow one benzene molecule to displace another without complete dissociation of the capsule. This is consistent with participation of an entering benzene in the rate-determining step. The intermediate structures are shown: in Figure 3.19(A), the capsule is held together by six hydrogen bonds; in Figure 3.19(B), two adjacent walls are open and five hydrogen bonds remain intact. Either of these can be regarded as a representative example of small guest exchange.

Large Molecules

For large guest exchange, we studied the supramolecular substitution of encapsulated 4,4'-dimethylbiphenyl (**2**) with incoming guest 4,4'-dimethylstilbene (**3**); the reaction is practically irreversible, since **3** is favored in equilibrium studies. The initial rates as a function

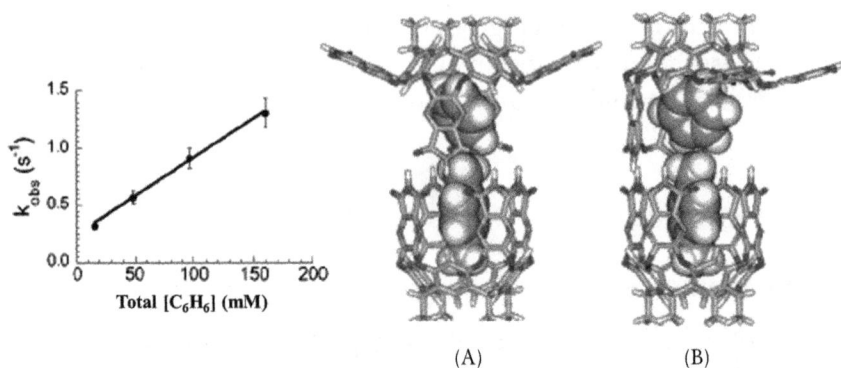

Figure 3.19 *Left*: Rates of benzene exchange as a function of its concentration; error bars are ±10%. *Right*: Opening of opposite walls (A) or adjacent walls (B) allows the substitution of guest benzene by solvent benzene to take place. [Reprinted with permission from *J Am Chem Soc* **124**, 8780–8781. Copyright 2002, American Chemical Society.]

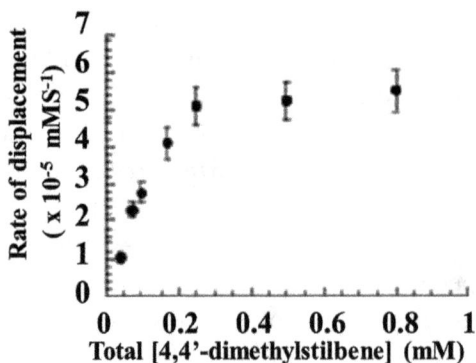

Figure 3.20 *Left*: Rates of dimethyl biphenyl displacement as a function of dimethyl stilbene concentration; error bars are ±10%. At 0.3 mM and above, the substitution process goes from S_N2- to S_N1-like. [Reprinted with permission from *J Am Chem Soc* **124**, 8780–8781. Copyright 2002, American Chemical Society.]

of concentration are shown in Figure 3.20, where the break in the plot indicates a change in the rate-determining step. The mechanism requires an intermediate and is similar to that proposed for the softball capsule.[40] Displacement of the resident guest by solvent leaves an intermediate that can either reclaim **2** or bind **3**. At high concentrations of the incoming guest, every intermediate leads to substitution, and the kinetics show saturation. An inverse dependence of the exchange rate on the concentration of outgoing guest **2** in solution was observed, and it further supports this mechanism. Neither complete dissociation of the capsule nor intermediacy of an "empty" capsule is likely.

Forcing Exchange

The exchange rate of encapsulated *trans*-4,4-dimethyl-azobenzene with *n*-tridecane as an incoming guest was determined by Henry Dube.[41] Even when 30 equivalents of *n*-tridecane compete for the capsule with a single azobenzene, only 19% of the azobenzene is replaced at equilibrium, which is reached only after a day. The exchange rates determined by fluorescence methods are consistent with these findings.[42]

The exchange rate is very slow and the azobenzene is favored. However, after irradiation at 365 nm wavelength for 50 min at 20°C the azobenzene is completely replaced by *n*-tridecane, and *cis*-azobenzene is present in solution (Figure 3.21).

The light-induced replacement of azobenzene by *n*-tridecane is completely reversible. After heating the sample to 160°C for 2 min, the initial state is restored, with the azobenzene as the only guest. The cycle can be repeated indefinitely. The size of the incoming molecule can also vary, as completely parallel results were obtained for the smaller benzamide or benzoic acid as the incoming guest. After irradiation, the respective hydrogen-bonded dimers are the only guests present.

For entropic reasons, a single large guest generally replaces two smaller occupants,[43] but the forced photoisomerization can reverse the trend. Again, brief heating restores the initial state (Figure 3.21). This cycle can also be repeated many times.

The mechanism involves *trans–cis* isomerization of the azobenzene inside the capsule, which forces out at least one wall. The *cis*

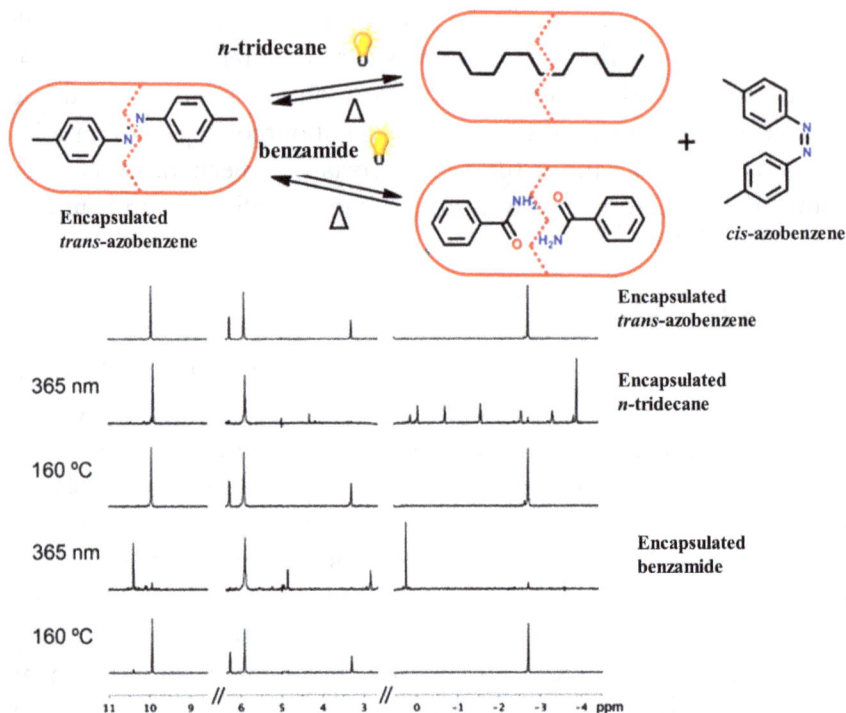

Figure 3.21   Encapsulated azobenzene responds to irradiation and breaks out of the capsule. Other guests rush in to fill the void. Brief heating resets the system to its resting state.

isomer cannot be accommodated and is rapidly flushed out by the incoming guest (Figure 3.22). Guest exchange in deep cavitands[44] involves related vase-to-kite conformational changes of the walls but Diederich *et al.* have recently established that only two walls need to undergo this motion to enable guest exchange.[45]

Protic Solvents

A means of releasing guests is through destruction of the seam of hydrogen bonds that hold the capsule together by the addition of competitive solvents like alcohols. The lifetime of a capsule occupied by a small guest in mesitylene is ~0.5 s, a convenient rate for NMR studies, and Toru Amaya first examined exchange rates as a

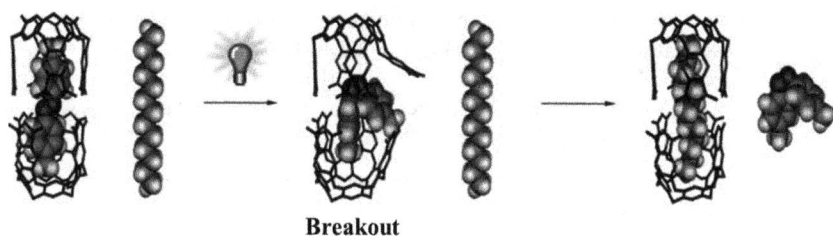

Breakout

Figure 3.22 The *cis* azobenzene is too wide to fit and forces out a wall. The C13 rushes in to replace the azobenzene. [Reprinted with permission from *Angew Chem Int Ed Eng* **49**, 3192–3195. Copyright 2010, Wiley-VCH, Weinheim.]

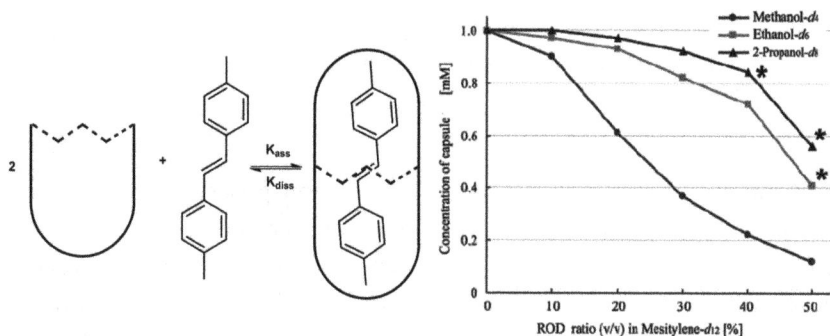

Figure 3.23 The stability of the capsule depends on the guest and the response to hydroxylic solvents. Methanol is more disruptive than isopropanol. [Reprinted with permission from *J Am Chem Soc* **126**, 14149–14156. Copyright 2004, American Chemical Society.])

function of added CD_3OD.[46] His results yet again showed that a capsule with multiple guests is not as stable as a well-filled capsule with a single guest. With 4,4'-dimethyl stilbene as the guest, some (12%) capsule complex persisted at 50% CD_3OD in the solvent mesitylene-d_{12}. Studies with ethanol-d_6 and 2-propanol-d_8 showed progressively increased stability (Figure 3.23). In these cases, equilibrium constants $K = $ [capsule]/[cavitand]^2[guest] could be determined, and at 30% protic solvents the free energies of capsule formation were: methanol -5.5, ethanol -7.4 and 2-propanol -8.5 kcal/mol.

The observation of a free cavitand, presumably occupied by solvent(s), presented a rare opportunity to obtain both thermodynamic

Table 3.3 Thermodynamic Parameters in the Association–Dissociation Equilibrium ($[\,1]_{total} = 2.0\,mM$; $[8]_{total} = 25\,mM$; Methanol-d_4 + Mesitylene– $d_{12} = 600\,\mu L$), ΔH (kcal/mol) and ΔS (cal/mol)]. [Reprinted with Permission from *J Am Chem Soc* 126, 14149–14156. Copyright 2004, American Chemical Society.]

Methanol-d_4 (v/v in mesitylene-d_{12})	ΔH^{a}	ΔS^{b}
10%	6.4	49
20%	2.7	30
30%	1.5	23
40%	c	c
50%	c	c

[a] ΔH values were obtained by van 't Hoff plots.
[b] $\Delta G = \Delta H - T\Delta S$.

and kinetic parameters for the encapsulation process. Because they point to differences between hydroxylic and noncompeting solvents, we dwell on them here. A van 't Hoff plot showed that the process was entropy-driven in methanol/mesitylene (Table 3.3). The liberation of many methanol solvents from the upper rim of the cavitand is the likely cause. The hydrogen-bonding inventory[47] is approximately 0, and is reflected in the modest enthalpy changes.

The rates of capsule dissociation and recombination were determined using EXSY NMR spectroscopy, as intense EXSY cross-peaks were observed between the capsule resonances and the signals of the dissociated cavitand. This gave the rates reported in Table 3.3. Cross-peaks were also observed between the stilbene guest inside and outside the capsule and allowed the determination of guest exit (k_{out}) and entrance (k_{in}). There was no significant difference between the observed rate for guest exit and capsule dissociation ($k_{out} = 0.17\,s^{-1}$; $k_{diss} = 0.16 s^{-1}$) or between guest entrance and capsule association k_{in} and k_{ass} ($k_{in} = 4.8 \times 10^4\,M^{-2}s^{-1}$; $k_{ass} = 4.2 \times 10^4$ $M^{-2}s^{-1}$). In short, complete dissociation of the capsule is required to exchange this large guest in the presence of these protic solvents. This is in contrast to the case in noncompeting solvents where guest exchange is invariably faster than capsule dissociation.

Other Cylindrical Capsules

A Thiourea Capsule

The breathing motions of the hydrogen-bonded seam widen any capsule's dimensions, but a permanent way of making the capsule wider was conceived by Ali Asadi: he decided to replace the four ureas (benzimidazolones) with thioureas. This substitution was not expected to change the dimensions of the capsule very much, since the increased acidity of the thioureas would make them better hydrogen bond donors, but the lower electronegativity of Sulfur vs Oxygen would make them poorer acceptors. These considerations are the underpinnings of the widespread use of thioureas in organocatalysis[48] — they selectively bind and enhance the electrophilicity of carbonyls. Modeling did show a promising increased girth of the thiourea capsule with respect to the urea. They are superimposed in Figure 3.24.

The capsule was prepared uneventfully, and we used NMR and the usual series of *n*-alkanes as probes of the space. For the shorter alkane C10–C13, signals in the NMR spectrum showed very much the same patterns as they did with the urea capsule.[49] However, the methyl signals (−3.5 ppm) were not as upfield shifted as they were in the tetraimide (−4.0 ppm). This suggests that the methyls are, on average, further away from the thiourea capsule's walls. But a qualitative change appeared in the spectrum of the longer C14. Its methyl signal moved *downfield*, which shows a change of the methyl's environment out of the depths of the cavity. Even C15 is encapsulated — an

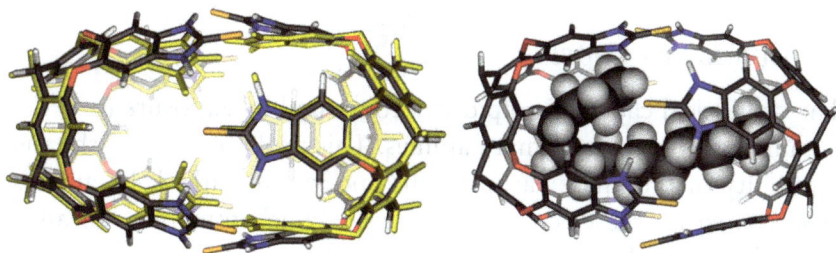

Figure 3.24   *Left*: Overlay of the original capsule and the thiourea analog. *Right*: Folded alkanes can be accommodated in the wider thiourea capsule.

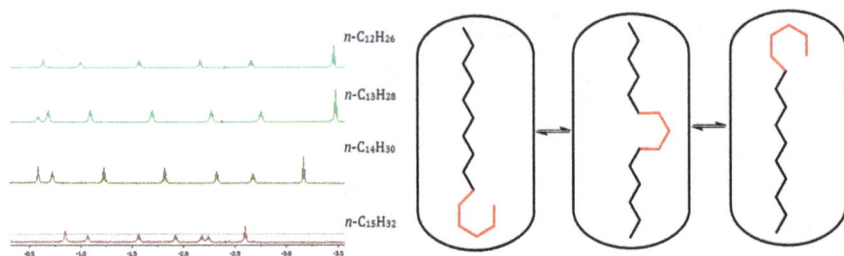

Figure 3.25 *Left*: Gradual downfield shifts of the methyl group indicate folding of the alkane. *Right*: Cartoon depiction of the motion of the alkane in the capsule. [Reprinted with permission from *J Am Chem Soc* 133, 10682-10684. Copyright 2011, American Chemical Society.]

unprecedented guest length for such containers — and its terminal methyl signal has moved even further downfield (Figure 3.25).

The seven NMR signals report an average for the proton environments. At any given time the two methyl groups are in different areas of the capsule: one is deep in the cavitand, while the other is out and near the walls. In short, the guest molecule is starting to fold. The methyls exchange magnetic environments rapidly on the NMR timescale, and the cartoon shows a means by which this can be accomplished. The folded conformation needs a wider capsule, which is apparently provided by the thiourea hydrogen-bonding characteristics. These folded alkane conformations are unknown solution and are even rare in enzyme crystal structures.[50] They do appear in systems that are much more rigid, such as metal coordination capsules[51] and covalent containers.[52]

de Mendoza's Capsule

The cylindrical capsule is rapidly decomposed by exposure to strong organic bases such as primary amines. This probably reflects the susceptibility of the imide carbonyls to nucleophilic attack by amines, exacerbated by the electron-deficient pyrazine aromatics. An alternative hydrogen-bonding pattern, but one that leads to a capsule of the same size, was introduced by de Mendoza.[53] He used the benzimidazolone skeleton of Cram's velcrand and, like our approach,

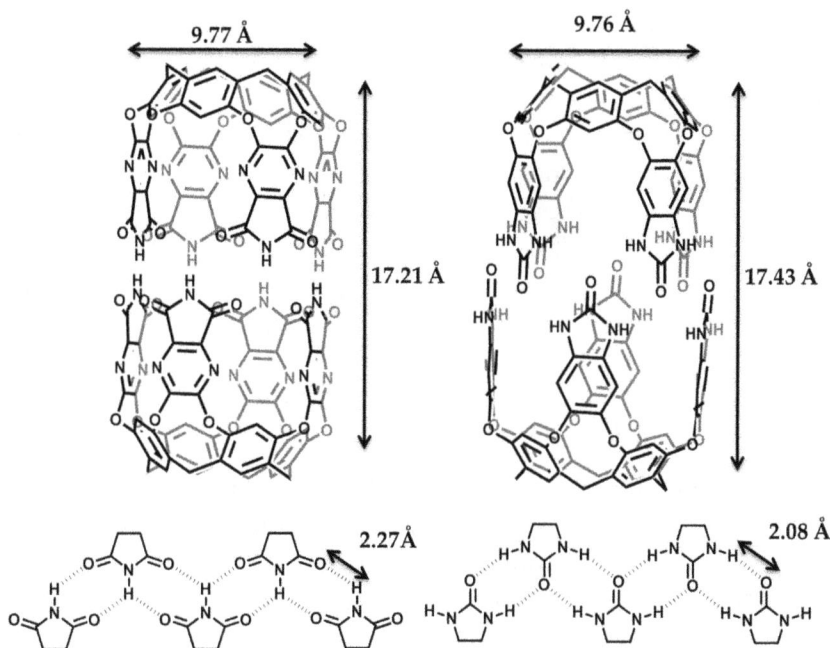

Figure 3.26   Hydrogen-bonding patterns of the original capsule and its urea analog; 16 hydrogen bonds hold the urea capsule together.

simply removed all the N- and C-methyl groups. The upper rim of this cavitand also offers a self-complementary hydrogen bond array, and 24 strong hydrogen bonds can be formed (Figure 3.26).

The benzimidazolones are heterocycles stable to nucleophiles, acids and bases. While the synthesis of the cavitand module with long chain hydrocarbon "feet" was uneventful, the behavior of the molecules in organic solvent was unexpected. The assembly process did not stop with simple dimerization. Instead, aggregation occurred to give what was characterized as large reverse vesicles. However, this aggregation was sensitive to the nature of the "feet." A shorter chain incorporating a *cis* double bond was found to prevent aggregate formation, and reasonable NMR spectra indicated the compound to be a dimeric capsule — a conclusion confirmed by MALDI mass spectrometry and vapor pressure osmometry. Encapsulation complexes were obtained and characterized with carboxylic acid guests. Surprisingly, given the dimensions and shape, which were almost exactly like those of the tetraimide cylindrical capsule, dicyclohexylcarbodiimide was not taken up by the new capsule.

Two years later, Choi and coworkers[54] reported the synthesis of the same capsule and showed that problems of aggregation could be overcome by loading the guests at elevated temperatures in mesitylene. A typical guest, 4,4'-dimethyl benzanilide was encapsulated in this way; after formation of the complex, evaporation of the solvent gives a much more soluble assembly.

Water Soluble Capsules

The stronger and more numerous hydrogen bonds in the benzimida-zolone capsule made it a reasonable candidate for stability in a competitive solvent such as water. But how to make the molecule dissolve in this solvent? Kang-da Zhang and Dariush Ajami decided to incorporate ionic groups on the periphery in order to impart solubility, but avoided the use of ionizable functions, i.e. those functions that would have a pH sensitivity to their charges.

Accordingly, with Jesse Gavette they prepared the chloro compound without event, then treated it with pyridine as the solvent to give the tetrapyridinium cavitand. This molecule had reasonable solubility in water (D_2O) at millimolar concentrations and showed the typical binding behavior of cavitand. For example, hydrophobic groups such as octanoic acid and its derivatives were bound inside and high affinity was shown for the octanoyl group of the gastric peptide ghrelin.[55] Sonication with longer hydrocarbons gave complexes that showed all of the symmetry and NMR spectra expected for a capsule in aqueous solution (Figure 3.27). The maximum upfield signal of the terminal methyls at −3.8 ppm indicated a shift $\Delta\delta$ of −4.6 ppm, a value close to that in the tetraimide ($\Delta\delta$ −4.8 ppm). But how could one tell if the hydrogen-bonded seam was intact? After all, there was much evidence from the Gibb[56] and Ramamurthy[57] groups that simple hydrophobic effects could drive two cavitands to form a capsule around a long guest, even when no hydrogen bonding was possible.

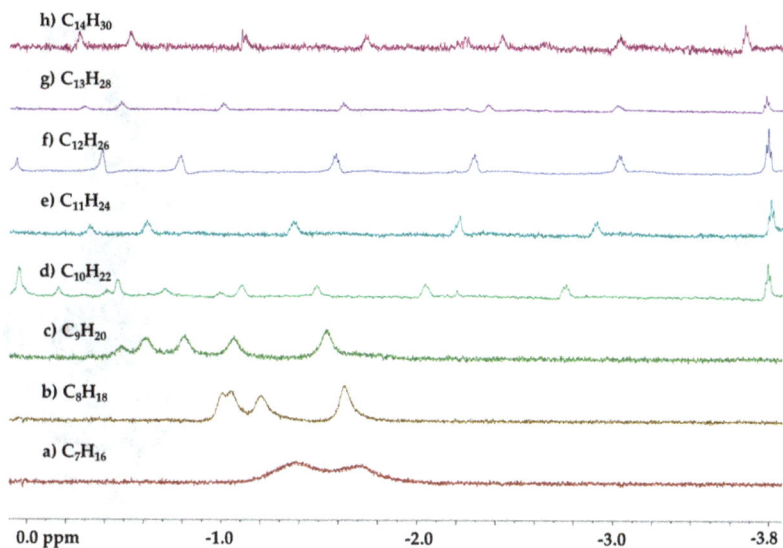

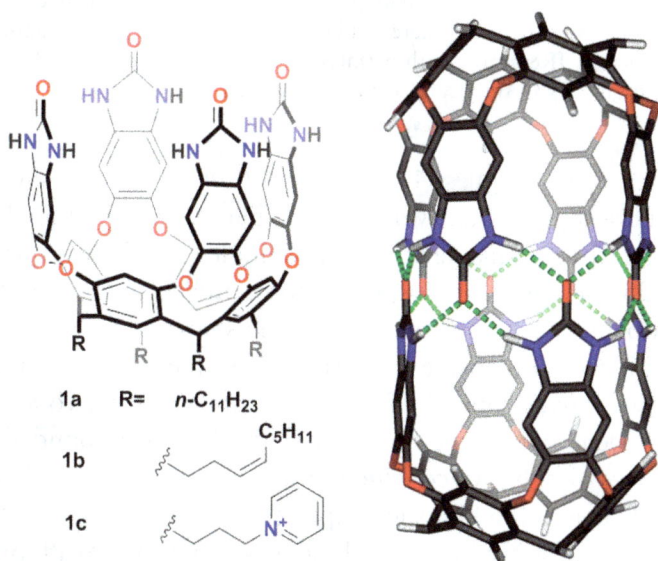

Figure 3.27 *Top*: Partial ^{1}H NMR (600 MHz, D$_2$O, 298 K) spectra of the complexes formed between host **1c** (0.5 mM) and normal alkanes. *Bottom*: The formula of the cavitands and their dimeric capsule with peripheral groups removed. [Reproduced from *Organ Bio Chem* **5**, 79, with permission from The Royal Society of Chemistry.]

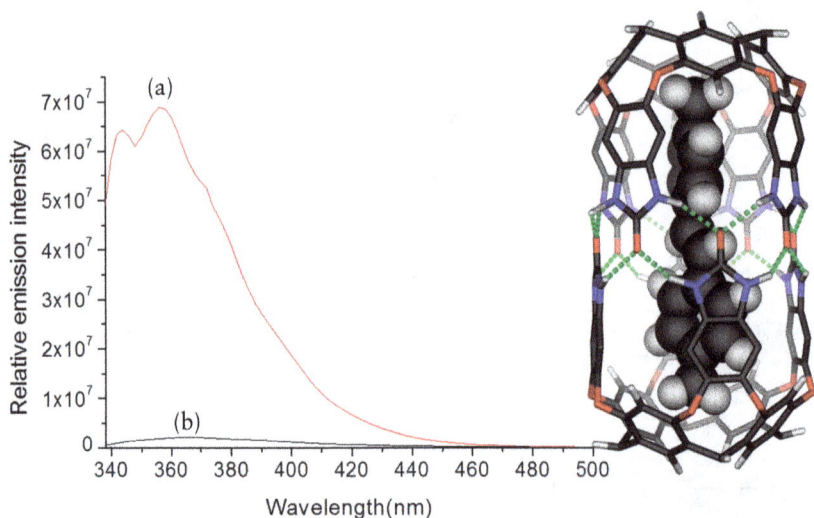

Figure 3.28 *Left*: Quenching of fluorescence occurs for encapsulated stilbenes. *Right*: The ground state of encapsulated stilbene involves a twist, and aromatic planes cannot be coplanar. [Reprinted with permission from *J Am Chem Soc* **135**: 18064–18066. Copyright 2013, American Chemical Society.])

Zhang used to the guest 4,4′-dimethyl-*trans*-stilbene to show that the capsule's shape space was fixed. That is to say, the fluorescence of the stilbenes inside this water-soluble capsule was quenched (Figure 3.28), as would be expected for a molecule that could not reach a coplanar ground state.[58] Since the shape of the space inside the benzimidazalone capsule is the same as inside the tetraimide capsule (the two halves rotated 45° from alignment), the two aromatic rings of stilbene cannot be coplanar if the hydrogen-bonded seam is intact. Accordingly, we concluded that the hydrogen-bonding was maintained. This is the first such capsule that exists in aqueous solution and even guest binding in such polar solvents as hexafluoro-isopropanol has been demonstrated.

The loosened belt of the thiourea capsule and its ability to accommodate folded alkanes prompted us to examine the longer alkanes in the water-soluble capsule. Its breathing motions should be enhanced by nearby solvents, and even the insertion of water(s) into the seam could be possible. Gavette and Zhang sonicated the

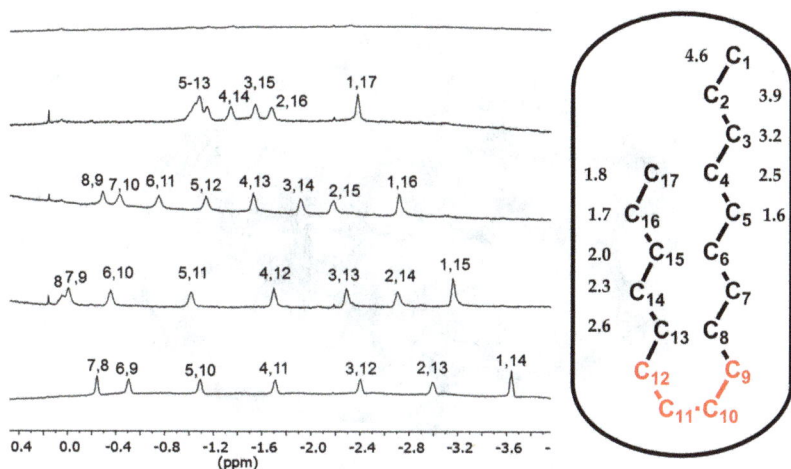

Figure 3.29 *Left*: Downfield shifts of methyl groups indicate folding of alkanes in the water-soluble capsule. *Right*: Structure of **C17** inside the capsule. [Reproduced from *Organ Bio Chem* **5**, 79, with permission from The Royal Society of Chemistry.]

cavitand with long and unlikely guests C15–C18 and were rewarded by encapsulation. The NMR signals of the guests move monotonically downfield in accord with increasingly folded conformations for C15–C17 (Figure 3.29). A model and a structure of C17 inside are shown (Figure 3.30); they are consistent with the NMR spectra.[59] Even traces of C18 can be detected, which must be neatly folded in half to fit inside.

The combination of NICS calculations on the capsules and the flexibility of *n*-alkanes provide useful probes of the shape, capacity and chemical surfaces of synthetic containers. The alkanes assume shapes complementary to the container, and the conformations can be mapped out with some confidence. These shapes are often unknown for alkanes in bulk solution or the gas phase, so the small spaces present a transient complementary phase.

Capsule Exchange

The association constants for capsule dimerization in mesitylene are beyond the reach of NMR titrations, typically $>10^5$ M^{-1}, so at

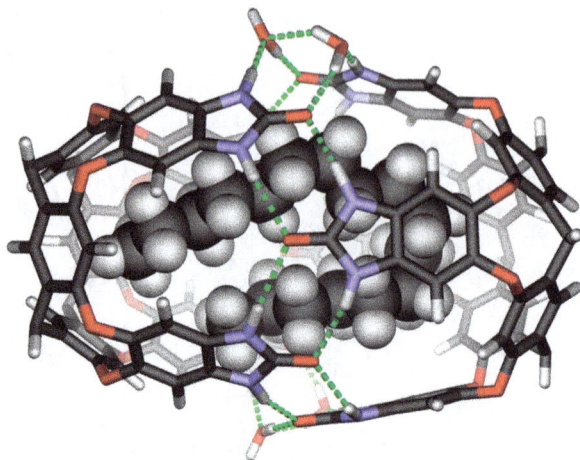

Figure 3.30 Energy-minimized structure of the folding alkane indicates that the seam of hydrogen bonds is compromised by solvent water molecules.

millimolar concentrations no cavitand (half-capsule) is observed. The exchange rate of cavitands between capsules is obviously a matter of guest size. For small guests such as benzene coencapsulated with *p*-xylene, guest exchange occurs within seconds and without capsule dissociation. For large guests, exchange rates were slow and took days to reach equilibrium. A chiral version of the capsule was used in an attempt to monitor exchange of the capsule halves by ^{1}H NMR spectroscopy. A solution of the chiral capsule containing 4,4'-dimethylbiphenyl was mixed with a solution of an achiral capsule having the same guest, and an equilibrium mixture was obtained with a half-life of approximately 10 min.[60] This equilibration of the capsules was found to be independent of the incoming guest concentration. But similar experiments attempted with 25 other guests did not show separate signals for the hybrid capsule versus the homodimers, so we turned to other, more general methods.

Fluorescence resonance energy transfer (FRET) is a technique which is widely employed in biological systems for the study of assembly and dynamic processes in real-time.[61] In an earlier chapter, we described the application of FRET in the study of calixarenes and we used it on the cylindrical capsule as well. Elizabeth Barrett and

Trevor Dale prepared the cavitands bearing a FRET dye pair following well-trodden paths (Figure 3.31).[62] They each formed capsules with suitable guests, and the NMR spectra of the encapsulated guests showed only a single set of resonances, even though two isomeric capsules are present.

A number of guest types were studied, including rigid biphenyls, less rigid benzanilides and flexible *n*-alkanes. For each type, pairwise competition experiments were conducted to establish the relative binding affinities (K_{rel}) of the guests for the capsule. These are reported in Tables 3.4 and 3.5, along with the rates and other parameters. A large excess of the guest was employed and diluted in mesitylene to the initial concentration of 250 nM for each capsule.

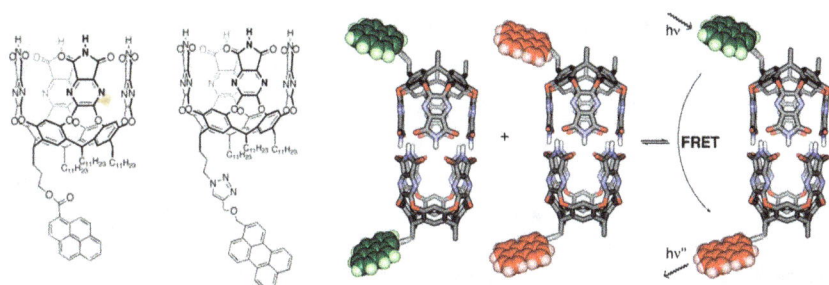

Figure 3.31   *Left*: Cavitands labeled with fluorescent dyes. *Right*: Exchange of capsule components leads to FRET which be monitored in real-time.

Table 3.4   Measured Rates (*k*) and Half-lives for the Exchange of Capsule Subunits with Substituted Benzanilide Guests 9–13 in Mestylene. Relative Binding Affinities (K_{rel}) of the capsule for the Guests in d_{12}-Mesitylene, and Guest Lengths.

Guest	$K(s^{-1})^a$	Half-life^a	K_{rel}	Length (Å)^b
9	2.0×10^{-2}	34 s	<0.1	11.4
10	6.1×10^{-4}	18 min	2.7	12.3
11	2.0×10^{-4}	58 min	5.3	12.3
12	5.4×10^{-6}	36 h	100	13.2
13	1.1×10^{-4}	108 min	1.8	14.7

Table 3.5   Rates and Half-lives for the Exchange of Capsule Subunits with Alkane Guests in Mesitylene and Relative Binding Affinities of the Capsule Fir the Guests in d_{12}-Mesitylene.

Guest	$k(s^{-1})^a$	Half-life[a]	K_{rel}
C_9H_{20}	1.6×10^{-3}	7 min	0.3
$C_{10}H_{22}$	5.4×10^{-6}	36 h	16.9
$C_{11}H_{24}$	$\gg 2 \times 10^{-6}$	$\gg 100$ h	100
$C_{12}H_{26}$	3.8×10^{-6}	51 h	24.4
$C_{13}H_{28}$	8.1×10^{-6}	24 h	1.0
$C_{14}H_{30}$	6.8×10^{-6}	28 h	13.2

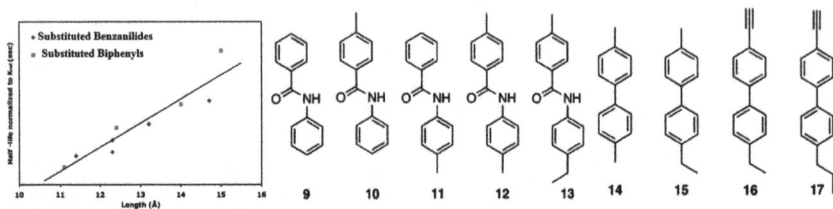

Figure 3.32   Correlation of capsule exchange rates with guest length. [Reprinted with permission from *J Am Chem Soc* **129**: 8818–8824. Copyright 2007, American Chemical Society.]

For the rigid guests there exists a clear correlation of the capsule exchange rate with the guest length (Figure 3.32). This speaks for an exchange mechanism that exposes large openings — perhaps even a kite conformation — in the cavitand subunits. Since there are unlikely to be empty containers, a second guest must be entering the cavitand departing the capsule.

For the flexible alkane guest there is a somewhat weaker correlation between exchange rates and binding affinity. The fully extended C11 is the best guest and its capsule is the most reluctant to exchange halves, but even the weakly bound and compressed C14 exchanges slowly. The conformations of these deep cavitands were originally observed as either vase or kite, but recent constructs by Diederich,[63] who prepared a cavitand with covalently fixed opposite walls, reveal the existence of other shapes available along the exchange process pathway.

References

1. Högberg AGS. (1980) Two stereoisomeric macrocyclic resorcinol-acetaldehyde condensation products. *J Org Chem* **45**, 4498.
2. Moran JR, Ericson JL, Dalcanale E, *et al.* (1991) Vases and kites as cavitands. *J Am Chem Soc* **113**, 5707–5714.
3. Dalcanale E, Soncini P, Bacchilega G, Ugozzoli F. (1989) Selective complexation of neutral molecules in organic solvents. Host–guest complexes and cavitates between cavitands and aromatic compounds. *J Chem Soc Chem Commun* 500–502.
4. Cram DJ, Choi HJ, Bryant JA, Knobler CB. (1992) Host–guest complexation. 62. Solvophobic and entropic driving forces for forming velcraplexes, which are 4-fold, lock–key dimers in organic media. *J Am Chem Soc* **114**(20), 7748–7765.
5. Soncini P, Bonsignore S, Dalcanale E, Ugozzoli F. (1992) Cavitands as versatile molecular receptors. *J Org Chem* **57**, 4608–4612.
6. Moran JR, Karbach S, Cram DJ. (1982) Cavitands: synthetic molecular vessels. *J Am Chem Soc* **104**, 5826–5828.
7. Ajami D, Rebek Jr J. (2013) Unexpected consequences of methyl substitutions in supramolecular chemistry. *Supramol Chem* **25**, 574–580.
8. Heinz T, Rudkevich D, Rebek Jr J. (1998) Pairwise selection of guests in a cylindrical molecular capsule of nanometre dimensions. *Nature* **394**, 764–766.
9. Körner SK, Tucci FC, Rudkevich DM, Heinz T, Rebek Jr J. (2000) A self-assembled cylindrical capsule: supramolecular phenomena through encapsulation. *Chem Eur J* **6**, 187–195.
10. Heinz T, Rudiveich DM, Rebek Jr J. (1999) Molecular recognition within a self-assembled cylindrical host. *Angew Chem Int Ed Engl* **38**, 1136–1139.
11. Hayashida O, Sebo L, Rebek Jr J. (2002) Molecular discrimination of N-protected amino acid esters by a self-assembled cylindrical capsule: spectroscopic and computational studies. *J Org Chem* **67**, 8291–8298.
12. Scarso A, Trembleau L, Rebek Jr J. (2003) Encapsulation induces helical folding of alkanes. *Angew Chem Intl Ed Engl* **115**, 5657–5660.
13. Schleyer PvR, Maerker C, Dransfeld A, *et al.* (1996) Nucleus-independent chemical shifts: a simple and efficient aromaticity probe. *J Am Chem Soc* **118**, 6317–6318.
14. Eliel E, Wilen SH. (1994) Conformation of Acyclic Molecules (Chap. 10), in *Stereochemistry of Organic Compounds*. Wiley, New York.
15. Scarso A, Trembleau L, Rebek Jr J. (2004) Helical folding of alkanes in a self-assembled, cylindrical capsule. *J Am Chem Soc* **126**, 13512–13518.
16. Jiang W, Ajami D, Rebek Jr J. (2012) Alkane lengths determine encapsulation rates and equilibria. *J Am Chem Soc* **134**, 8070–8073.

17. Ajami D, Rebek Jr J. (2009) Compressed alkanes in reversible encapsulation complexes. *Nature Chem* **1**, 87–90.
18. Siering C, Torang J, Kruse H, *et al.* (2010) Enantioselective helical folding inside a self-assembled, cylindrical capsule. *Chem Commun* **46**(10), 1625–1627.
19. Wanjari PP, Sangwai AV, Ashbaugh HS. (2012) Confinement induced conformational changes in *n*-alkanes sequestered within a narrow carbon nanotube. *Phys Chem Chem Phys* **14**, 2702–2709.
20. Ajami D, Iwasawa T, Rebek Jr J, (2003) Experimental and computational probes of the space in a self-assembled capsule. *Proc Natl Acad Sci USA* **103**, 8934–8936.
21. Shivanyuk A, Scarso A, Rebek Jr J (2003) Coencapsulation of large and small hydrocarbons. *Chem Commun* **11**, 1230–1231.
22. Rechavi D, Scarso A, Rebek Jr J. (2004) Isotopomer encapsulation in a cylindrical molecular capsule — a probe for understanding non-covalent isotope effects on a molecular level. *J Am Chem Soc* **126**, 7738–7739.
23. Zhao YL, Houk KN, Rechavi D, Scarso A, Rebek Jr J. (2004) Equilibrium isotope effects as a probe of nonbonding attractions. *J Am Chem Soc* **126**, 11428–11429.
24. Haino T, Fukuta K, Iwamoto H, Iwata S. (2009) Noncovalent isotope effect for guest encapsulation in self-assembled molecular capsules. *Chem Eur J* **15**, 13286–13290.
25. Liu Y, Warmuth RA. (2005) "Through-shell" binding isotope effect. *Angew Chem Int Ed* **44**, 7107–7110.
26. Laughrey ZR, Upton TG, Gibb BCA. (2006) Deuterated deep-cavity cavitand confirms the importance of C–H···X–R hydrogen bonds in guest binding. *Chem Commun* 970–972.
27. Mugridge JS, Bergman RG, Raymond KN. (2012) Equilibrium isotope effects on noncovalent interactions in a supramolecular host–guest system. *J Am Chem Soc* **134**, 2057–2066.
28. Hayashida O, Shivanyuk A, Rebek Jr J. (2002) Molecular encapsulation of anions in a neutral receptor. *Angew Chem Int Ed Engl* **41**, 3423–3426.
29. Scarso A, Onagi H, Rebek Jr J. (2004) Mechanically regulated rotation of a guest in a nanoscale host. *J Am Chem Soc* **126**, 12728–12729.
30. Eliel E, Wilen SH. (1994) Configuration and Conformation of Cyclic Molecules (Chap. 11), in *Stereochemistry of Organic Compounds*. Wiley, New York, pp. 690–700.
31. Fieser LF, Fieser M. (1959). *Steroids*. Reinhold, New York, p. 58.
32. Kumazawa K, Biradha K, Kusukawa T, *et al.* (2003) Multicomponent assembly of a pyrazine-pillared coordination cage that selectively binds planar guests by intercalation. *Angew Chem Int Ed Engl* **42**, 3909–3913.
33. Mills SG, Beak P. (1985) Solvent effects on keto–enol equilibria: tests of quantitative models. *J Org Chem* **50**, 1216–1224.
34. Jiang W, Tiefenbacher K, Ajami D, Rebek Jr J. (2012) Complexes within complexes: hydrogen bonding in capsules. *Chem Sci* **3**, 3022–3025.
35. Metrangolo P, Resnati G, eds. (2008) Halogen bonding: fundamentals and applications, in *Structure and Bonding 126*; Mingos DMP, series ed., Springer-Verlag, Berlin.

36. Cavallo G, Metrangolo P, Pilati T, *et al.* (2010) Halogen bonding: a general route in anion recognition and coordination. *Chem Soc Rev* **39**, 3772–3783.

37. Hauchecorne D, van der Veken BJ, Herrebout WA, Hansen PEA. (2011) ^{19}F NMR study of C–I$\cdots\pi$ halogen bonding. *Chem Phys* **381**, 5–10.

38. Sarwar MG Ajami D, Theodorakopoulos G, Petsalakis ID, Rebek Jr J. (2013) Amplified halogen bonding in a small space. *J Am Chem Soc* **135**, 13672–13675.

39. Craig SL, Lin S, Chen J, Rebek Jr J. (2002) Mechanism of single-molecule exchange in a cylindrical host capsule. *J Am Chem Soc* **124**, 8780–8781.

40. Santamaria J, Martin T, Hilmersson G, Craig SL, Rebek Jr J. (1999) Guest exchange in an encapsulation complex: a supramolecular substitution reaction. *Proc Natl Acad Sci USA* **96**, 8344–8347.

41. Dube H, Ajami D, Rebek Jr J. (2010) Photochemical control of reversible encapsulation. *Angew Chem Intl Ed* **49**, 3192–3195.

42. (a) Barrett ES, Dale TJ, Rebek Jr J. (2007) Self-assembly dynamics of a cylindrical capsule monitored by fluorescence resonance energy transfer. *J Am Chem Soc* **129**, 8818–8824; (b) Castellano RK, Craig SL, Nuckolls C., Rebek Jr J. (2000) Detection and mechanistic studies of multi-component assembly by fluorescence resonance energy transfer. *J Am Chem Soc* **122**, 7876–7822.

43. Kang J, Rebek Jr J. (1996) Entropically-driven binding in a self-assembling molecular capsule. *Nature* **382**, 239–241.

44. Cram DJ, Choi HJ, Bryant JA, Knobler CB. (1992) Solvophobic and entropic driving forces for forming velcraplexes, which are four-fold, lock–key dimers in organic media. *J Am Chem Soc* **114**, 7748–7765.

45. Gottschalk T, Jaun B, Diederich F. (2007) Container molecules with portals: reversibly switchable cycloalkane complexation. *Angew Chem Int Ed* **46**, 260–264.

46. Amaya T, Rebek Jr J. (2004) Hydrogen-bonded encapsulation complexes in protic solvents. *J Am Chem Soc* **126**, 14149–14156.

47. Fersht AR, Shi JP, Knill-Jones J, *et al.* (1985) Hydrogen bonding and biological specificity analysed by protein engineering. *Nature* **314**, 235–238.

48. Zhang Z, Schreiner PR. (2009) (Thio)urea organocatalysis — what can be learnt from anion recognition? *Chem Soc Rev* **38**(4), 1187–1198.

49. Asadi A, Ajami D, Rebek Jr J. (2011) Bent alkanes in a new thiourea-containing capsule. *J Am Chem Soc* **133**, 10682–10684.

50. Han GW, Lee JY, Song HK, *et al.* (2001) Structural basis of non-specific lipid binding in maize lipid-transfer protein complexes revealed by high-resolution X-ray crystallography. *J Mol Biol* **308**(2), 263–278.

51. Petina O, Rehder DE. Haupt TK, *et al.* (2011) Guests on different internal capsule sites exchange with each other and with the outside. *Angew Chem Int Ed* **50**, 410–414.

52. Ko YH, Kim Y, Kim H, Kim K. (2011) U-shaped conformation of alkyl chains bound to a synthetic receptor cucurbit[8]uril. *Chem Asian J* **6**, 652–657.

53. Ebbing MKH, Villa MJ, Valpuesta JM, *et al.* (2002) Resorcinarenes with 2-benzimidazolone bridges: self-aggregation, self-assembled dimeric capsules, and guest encapsulation. *Proc Natl Acad Sci USA* **99**, 4962–4966.

54. Choi HJ, Park YS, Cho CS, *et al.* (2004) Unusually stable molecular capsule formation of a tetraphenyleneurea cavitand. *Org Lett* **6**, 4431–4433.

55. Zhang KD, Ajami D, Gavette, JV, Rebek Jr J (2014) Complexation of alkyl groups and ghrelin in a deep, water-soluble cavitand. *Chem Commun* **50**, 4895–4897.

56. Liu S, Gibb BC. (2008) High-definition self-assemblies driven by the hydrophobic effect: synthesis and properties of a supramolecular nano-capsule. *Chem Commun* 3709–3716.

57. Kulasekharan R, Choudhury R, Prabhakar R, Ramamurthy V. (2011) Restricted rotation due to lack of free space within a capsule translates into product selectivity: photochemistry of cyclohexyl phenyl ketones within a water-soluble organic capsule. *Chem Commun* **47**, 2841–2843.

58. Ams MR, Ajami D, Craig SL, Yang JS, Rebek Jr J (2009) Control of stilbene conformation and fluorescence in self-assembled capsules. *Beilstein J Org Chem* **5**(79).

59. Gavette JV, Zhang K-d, Ajami D, Rebek Jr J. (2014) Folded alkyl chains in water-soluble capsules and cavitands. *Organ Bio Chem* **12**, 6561–6563.

60. Amaya T, Rebek Jr J (2004) Hydrogen-bonded encapsulation complexes in protic solvents. *J Am Chem Soc* **126**, 14149–14156.

61. Jameson DM, Croney JC, Moens PDJ. (2003) Fluorescence: basic concepts, practical aspects, and some anecdotes. *Methods Enzymol* **360**, 1–43.

62. Barrett ES, Dale TJ, Rebek Jr J. (2007) Self-assembly dynamics of a cylindrical capsule monitored by fluorescence resonance energy transfer. *J Am Chem Soc* **129**, 8818–8824.

63. Gottschalk T, Jaun B, Diederich F. (2007) Container molecules with portals: reversibly switchable cycloalkane complexation. *Angew Chem Int Ed* **46**, 260–264.

CHAPTER 4

Hexameric Capsules from Resorcinarenes and Pyrogallolarenes

In other chapters we have described how resorcinarenes have provided the structural curvature for several other capsules. But these molecules, and their pyrogallolarene cousins, have a capsular existence all of their own, and we explore them here. The recognition properties were the result of a convenient synthesis devised by Högberg,[1] and explored in the 1980s by Aoyama.[2] Reviews[3] cover these and related efforts up to 1996, but the Atwood and MacGillivray X-ray structure[4] of **1a** (Figure 4.1) showed an unprecedented assembly: a capsule resembling an inflated cube with six resorcinarenes as the sides and eight water molecules as the corners. The space inside was nearly 1400 Å^3. Mattay[5] showed that pyrogallol[4]arenes (**2**) also form hydrogen-bonded hexameric capsules in the solid state — and without the need for water. A number of smaller (dimeric) capsules with these modules were subsequently reported.[6] These systems are included in a recent review,[7] so they will not be covered here. Instead, we will limit our descriptions to the hexamers, and mostly to the work in the last decade.

One of the new structural developments was the introduction of resorcinarene capsules[8] derived from pyridines (**7**) by Mattay and coworkers.[9] These exist in chloroform solution as both the dimeric and hexameric capsules as determined by diffusion NMR,[10] a technique that has revolutionized the study of encapsulation.

While crystallography established the existence of hexamers in the solid state, the direct observation of guests inside the hexameric host capsule in solution was due to NMR studies. The resorcinarene, dissolved in *wet* chloroform with Hex$_4$N$^+$Br$^-$, showed

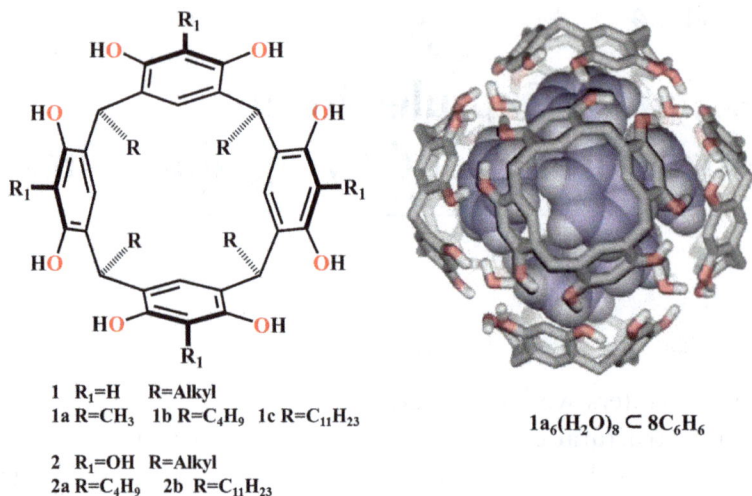

1 R_1=H   R=Alkyl
1a R=CH$_3$ 1b R=C$_4$H$_9$ 1c R=C$_{11}$H$_{23}$

1a$_6$(H$_2$O)$_8$ ⊂ 8C$_6$H$_6$

2 R_1=OH R=Alkyl
2a R=C$_4$H$_9$ 2b R=C$_{11}$H$_{23}$

Figure 4.1 *Left*: structures of resorcinarene **1** and pyrogallolarene **2** subunits. *Right*: The hexameric capsule of **1** with benzene guests; eight water molecules complete the array of hydrogen bonds.

separate sets of signals in the spectrum for free and bound salt.[11] Integration showed a 6:1 ratio for resorcinarene to ammonium salt, and the large upfield shifts of the bound salt left no doubt about its location in the capsule. The 24 aromatic rings provide the magnetic anisotropy and the polarizable π surfaces, while the positive charges on the convex surface of the ammonium guests complete the cation–π attractions. Smaller salts with their counterions and some solvent chloroform molecules were also coencapsulated, and even neutral molecules could be coaxed to go inside, if they filled the space properly.[12] Cohen has shown that even two molecules of the tetraethylammonium ion are readily accommodated by the capsule in CHCl$_3$.[13] The longest rigid guest to enter was 4-methylbiphenyl; the slightly longer 4,4-dimethylbiphenyl was not encapsulated. This defines some of the dimensions of the capsule that bears on the behavior of other, more flexible molecules inside. Using diffusion NMR, Cohen showed that large guests are not required to nucleate the formation of the capsules; even water-saturated solvents such as chloroform and benzene alone could be the occupants.[14] The encapsulated solvent molecules also exhibit upfield-shifted resonances in their NMR spectra. In short, lipophilic resorcin[4]arenes and pyrogallol[4]arenes spontaneously self-assemble into hexameric capsules.[15]

The behavior of solvents in the pyrogallol[4]arenes often differed from that in the resorcin[4]arenes: the chloroform molecules are in a statistical distribution of what seem like six slightly different environments. It is likely that these are different patterns of hydrogen bond arrangements of the capsule. The encapsulated CH_2Cl_2 is even more puzzling — two very different types of the *solvent* are present inside, each experiencing the same distribution of hydrogen bond seams.[16]

In wet, nondeuterated solvents the resorcinarene hexamer encapsulates what appear to be six and eight molecules of chloroform and benzene, respectively. The benzene peak is broadened (Figure 4.2(a)), indicating that exchange of benzene in and out of the capsule occurs on the NMR timescale. Or, as mentioned above, slightly different

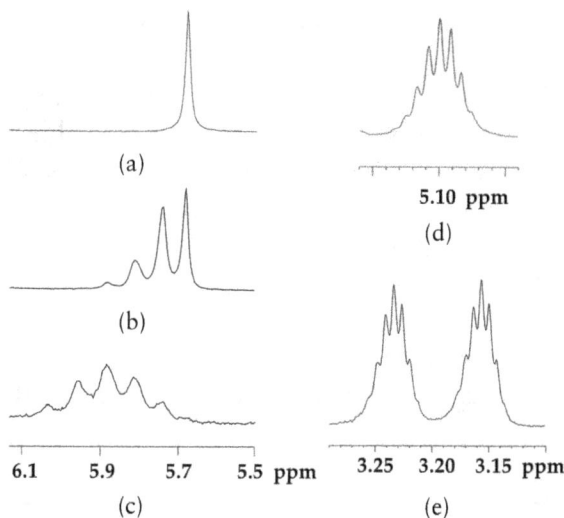

Figure 4.2 *Left*: The ^{1}H NMR signals (600 MHz, 298 K in a solution of resorcinarene **1b**) of encapsulated benzene alone (a) and with added $CHCl_3$ (b), (c). *Right*: The ^{1}H NMR signals (400 MHz, 298 K in a 20 mM $CHCl_3$ or CH_2Cl_2 solution, respectively) of pyrogallolarene **2b**. The encapsulated chloroforms show some seven slightly different signals (d); encapsulated CH_2Cl_2 molecules show this feature but two very different magnetic environments (e). [Reprinted with permission from *Org Lett* 8, 219–222. Copyright 2006, American Chemical Society.]

magnetic environments are experienced by the guest in isomeric capsules. On addition of aliquots of $CHCl_3$ the spectra show the coexistence of several capsules — the original with eight benzene guests, another with seven benzenes and one chloroform, etc. Eventually, a bell-shaped statistical distribution of all capsular species is generated. Chloroform molecules coencapsulated with benzene also show aromatic-solvent-induced shifts (ASIS). The close proximity of the coencapsulated molecules leads to exaggerated ASIS.

The Cohen group determined that eight water molecules are needed to form the hexameric capsules of resorcin[4]arene in solution,[17] but alcohols can also complete the seam of hydrogen bonds in the hexameric capsules.[18] When alcohols replace the water molecules in the structure of the hexameric capsules, they can be simply encapsulated, or part of the hexameric capsules from outside or encapsulated *and* part of the seam.[19] The possibilities are shown in Figure 4.3. Even tertiary alkylamines and quaternary alkylammonium interact with the capsules, from both the inside and the outside.[20]

Water molecules are not required for the pyrogallol[4]arene capsule's structure in solution,[21] and it can self-assemble in nonpolar organic solvents. With a suitable hydrocarbon guest as solvent, spontaneous formation of the hexameric capsule occurs even when the guest must contort itself to fill the space (Figure 4.4).

The ^{1}H NMR signals of the encapsulated hydrocarbon solvent molecules were obtained by Liam Palmer, and are shown in Figure 4.5.[22] Six molecules of the guest octane are taken in, one for each pyrogallolarene cavitand. The four NMR signals for

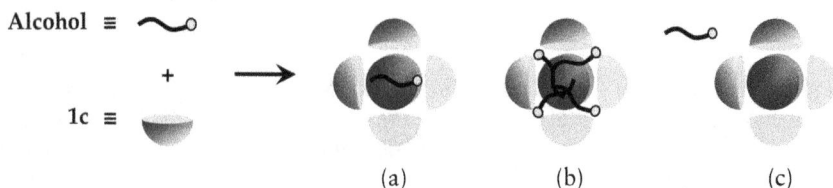

Figure 4.3 Possible sites (a)–(c) occupied by alcohols in a solution of the hexameric capsule of **1c**.

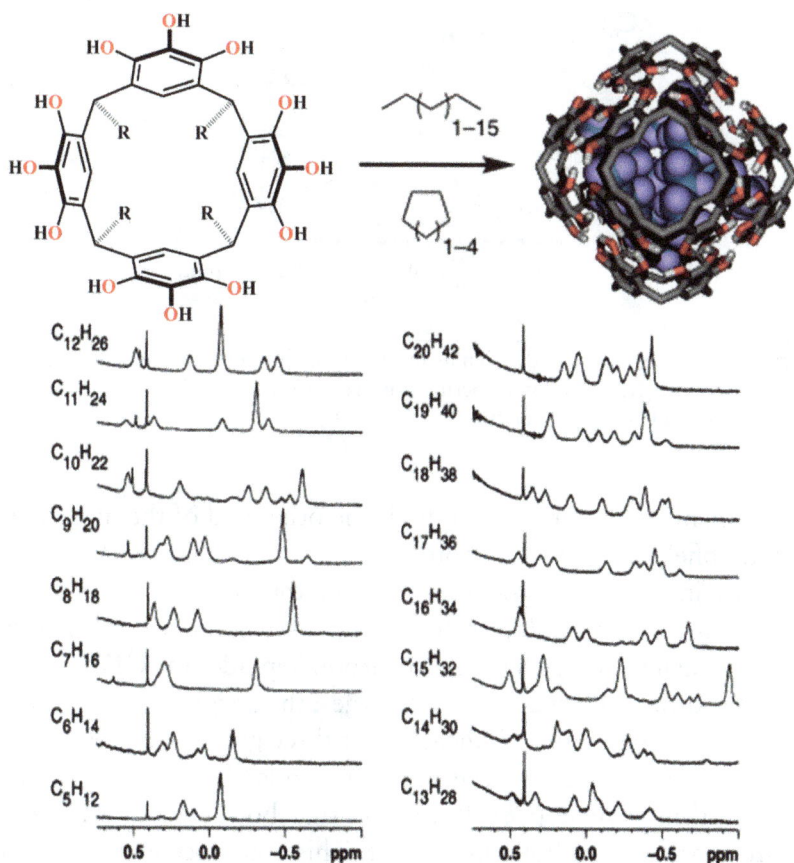

Figure 4.4 The hexameric pyrogallolarene encapsulates a variety of hydrocarbons. [Reprinted with permission from *Org Lett* **7**: 787–789. Copyright 2005, American Chemical Society.]

octane indicate that its two ends experience an averaged magnetic environment: the molecule is tumbling rapidly on the NMR timescale. At any given time, one methyl end group is near the pyrogallolarene and experiences a large upfield shift and the other methyl end is near the middle of the assembly, where only a small shift occurs. The tumbling exchanges these environments and the signal at -0.5 ppm is the average for the two environments. This signal represents an induced shift of -1.3 ppm (from where a methyl normally appears at $+0.8$ ppm). An estimate of the deeply buried methyl's observed

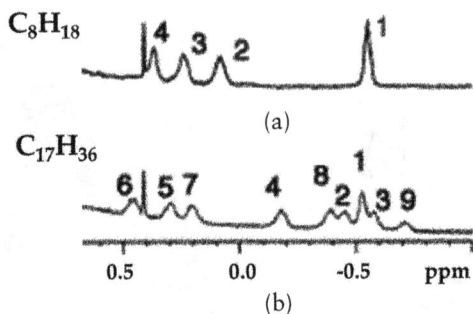

Figure 4.5 Assignments of the signals in the pyrogallol hexamer indicate that C17 is folded in half. [Reprinted with permission from *Org Lett* 7, 787–789. Copyright 2005, American Chemical Society.]

chemical shift would be −2.3 ppm if the other end of the molecule is shifted upfield by only −0.3 ppm.

With increased hydrocarbon length, the signal for the methyl groups moves *downfield*, as the group moves out of the shallow pocket. When the long chain hydrocarbon heptadecane C17 is encapsulated, a different signal pattern emerges; the assignments are shown in Figure 4.5. Integration indicates that three guests are present, and the largest upfield shifts occurred at its ends (C_1 and C_{17}) and in the central methylene (C_9). It appears that both ends and the middle are near a pyrogallolarene pocket. This could occur if (as above) the molecule tumbles rapidly *while it is folded in two*. The tumbling would be easier if the molecule is compressed, for example as in a helical coil. The upfield shifts for the signals of C_2 and C_3 (compared to octane) are consistent with such compression as it brings these atoms nearer the cavitand. The three guests would also have to move quickly between all six cavitands in this description, but there is no evidence for or against such a migration.

A series of alkylammonium salts examined by Masamichi Yamanaka and Alex Shivanyuk in the resorcin[4]arene **1c** (Figure 4.1) indicated that folding of alkyl groups is not unusual for encapsulation in these systems.[23] For example, the methyl group of the tetra-hexylammonium salt (Figure 4.6) enjoys the furthest upfield shift of $\Delta\delta$ − 2.2 ppm, a value in accord with the shifts of octane's methyl

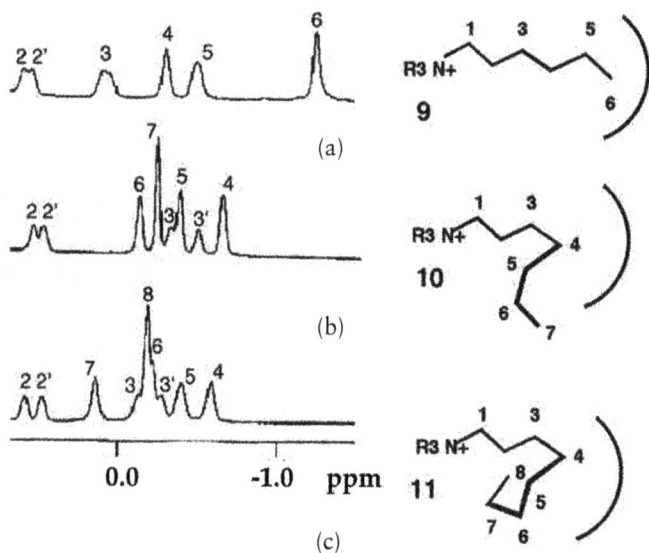

Figure 4.6 NMR signals for ammonium salts in the resorcinarene hexamer indicate buckling of the longer alkyl chains. [Reprinted with permission from *J Am Chem Soc* **126**, 2939–2934. Copyright 2004, American Chemical Society.]

groups encountered above ($\Delta\delta$ − 2.3 ppm). This value reflects the limit to which a (moving) methyl signal can be shifted in the cavity of a resorcinarene module of this hexamer. The signals of C_2 and C_3 are diastereotopic and speak for a desymmetrization such as folding near the N atom would cause. But the strongest evidence for folding is shown by the spectra of the longer heptyl and octyl groups. These show signs of "buckling" that pushes C_4 closest to the cavitand pocket.

A recent method of encapsulating large guests in the pyrogalloarene hexamer has been devised by Purse.[24] He employs solvent-free conditions using molten guest molecules as solvent. The capsules assemble during cooling, and give kinetically stable encapsulation complexes that are not formed in the presence of (competing) solvents. The complexes (Figure 4.7) are not thermodynamically stable with typical solvents and slowly convert to solvent-filled capsules.

A number of polycyclic aromatics could be encapsulated by this method. Paraffin wax may be used as a molten vehicle: the wax

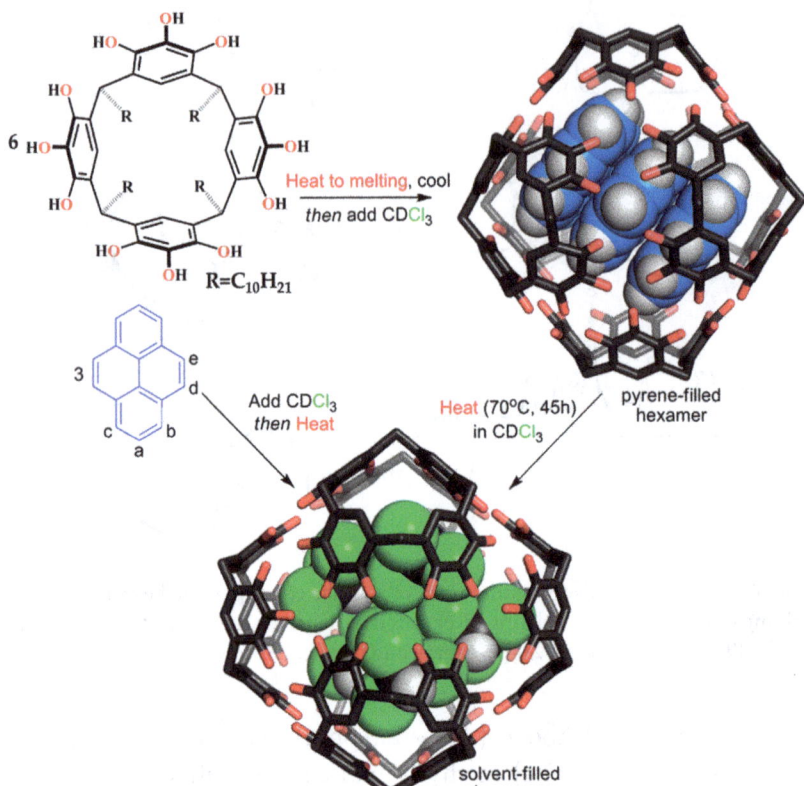

Figure 4.7   Solvent-free encapsulation of pyrene. The three pyrene guests are slowly replaced by added solvents such as CDCl$_3$. [Reprinted with permission from *J Am Chem Soc* 134: 15000–15009. Copyright 2012, American Chemical Society.]

molecules are too large to fit the cavity and so do not compete with the intended guests during capsule assembly.

Another innovation involving the hexamers concerns the fluorous phase. Shimizu has prepared a resorcinarene with fluorocarbon "feet" that assembles as usual but offers the advantages of solubility in fluorous media.[25]

A poorly understood difference between the hexameric capsules of resorcin[4]arenes and pyrogallo[4]arene is that the former can accommodate tertiary amines and tetraalkylammonium salts but the latter takes up only tertiary amines in chloroform solution. Protonation of tertiary amines in the pyrogallo[4]arene results in their ejection as

pK$_a$	11	9.2	8.8	8.4	7.0	6.1	5.2	4.6	3.8	1.7
degree of protonation (%)	80 ± 2	80 ± 2	77 ± 2	86 ± 3	83 ± 3	80 ± 2	53 ± 1	23 ± 2	np	np

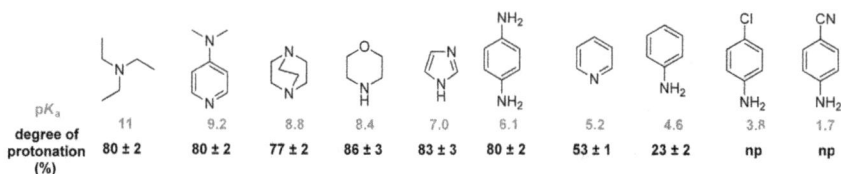

Figure 4.8 Amine guests of the resorcinarene hexamer in wet CDCl$_3$ and their protonation status inside. [Reprinted with permission from *J Am Chem Soc* **135**, 16213–16219. Copyright 2013, American Chemical Society.]

the ammonium salts.[26] DOSY NMR experiments established that the pyrogallo[4]arene capsule remains intact with CHCl$_3$ inside after the protonated amine departs.

Tiefenbacher investigated the physical properties of the resorcinarene hexamer and used NMR to determine the extent of protonation of amines inside (Figure 4.8). He found that the capsule is a reasonably strong Brønsted acid, with an effective pKa of approximately $5.5-6$.[27] The ammonium complexes are stabilized by cation–π interactions.

The capsules of resorcin[4]arenes are versatile receptors that take up a bizarre range of guest types. For example, sonication with L-phenylalanine or cyclohexylalanine in CDCl$_3$ results in encapsulation with a 1:6 guest–cavitand ratio, but other amino acids with hydrophobic side chains (Leu, Val, Tyr, Trp) are not taken up under the same conditions.[28]

Guest Exchange

We examined the exchange process for large tetraalkylammonium guests, with the usual premise that guest exchange would be faster than module exchange (Chapter 1).[26] This proved to be only partially true. Larger R$_4$N$^+$ guests leave more slowly than smaller guests and such a size dependence of the exchange (exit) rate is consistent with a "hole" created in an otherwise intact capsule. The activation free energies for exit as a function of R are: R = C$_3$ = 13.1; C$_4$ = 14.8; C$_5$ = 16.7; C$_7$ = 17.1 kcal mol^{-1}. At first glance, this could merely reflect

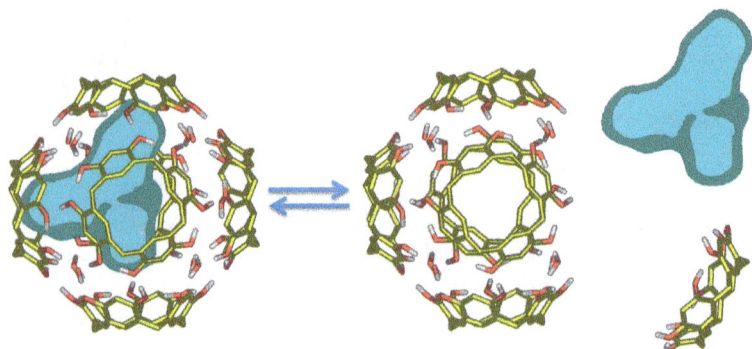

Figure 4.9   A tetraalkyl ammonium guest breaks out of the hexamer through an opening created by a departed resorcinarene module. [Reprinted with permission from *J Am Chem Soc* **126**, 2939–2934. Copyright 2004, American Chemical Society.]

tighter binding of the larger guests, but equilibrium studies and competition experiments show the opposite trend. For R_4N^+ the apparent association constants for encapsulation as a function of R are: $R = C_{3-5} > 10^4; C_6 = 1200; C_7 = 450; C_8 = 150\,M^{-1}$. Accordingly, it is necessary to propose an opening of limited size,[26] and a structure is suggested in Figure 4.9.

Self-Sorting

Resorcinarenes and pyrogallolarenes share a great deal in terms of size, shape and hydrogen-bonding surfaces that lead to their hexameric assemblies, and a reasonable question is: Do they self-sort? Self-assembling systems are generally believed to self-sort because the corrections that occur during the assembly process involve continuous distinctions between "self" and "non-self." Diffusion NMR techniques first showed that no evidence of scrambling of the components of the two respective hexameric capsules occurs (Figure 4.10). At the same time, scrambling does occur when two different resorcin[4]arenes or pyrogallol[4]arenes are used in the assembly.[21]

Elisabeth Barrett and Trevor Dale applied FRET techniques to the self-sorting issue. Resorcinarene hexamers bearing perylene or pyrene fluorophores readily exchanged their subunits when mixed

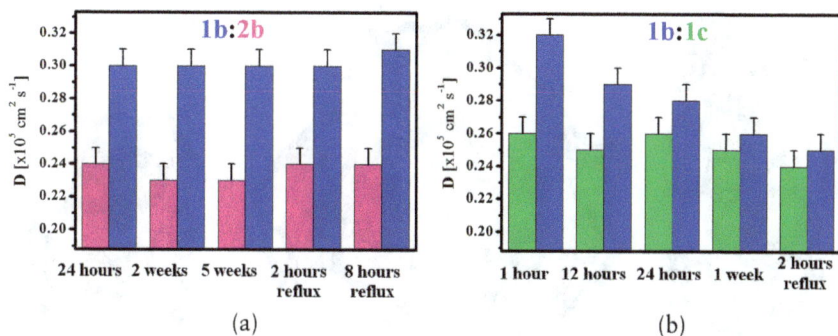

Figure 4.10 Diffusion coefficients of the peaks of **1b** (■), **1c** (■) and **2b** (■) in a mixture of (a) **1b:2b** and (b) **1b:1c** as a function of the time after preparation of the mixture and after 2 h and 8 h of reflux.[21] [Reprinted with permission from *J Am Chem Soc* **126**, 11556–11563. Copyright 2004, American Chemical Society.]

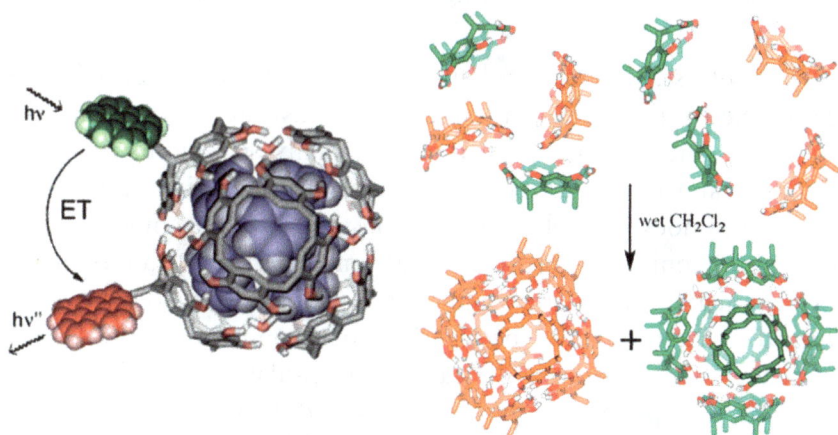

Figure 4.11 *Left*: Fluorescence resonance energy transfer takes place between two dyes if they are in the same assembly. *Right*: Strict self-sortingaddress = between resorcinarenes (green) and pyrogallolarenes (orange) occurs during their capsular assembly. [Reprinted with permission from *J Am Chem Soc* **130**, 2344–2350. Copyright 2008, American Chemical Society.]

(Figure 4.11).[29] A parallel study with hexamers of pyrogallolarenes[32b] showed the same scrambling behavior.[30] No evidence of exchange between a resorcinarene hexamer and a pyrogallolarene hexamer was found. Uncannily, a mixture of the modules was strictly self-sorting, even during the assembly process.[31]

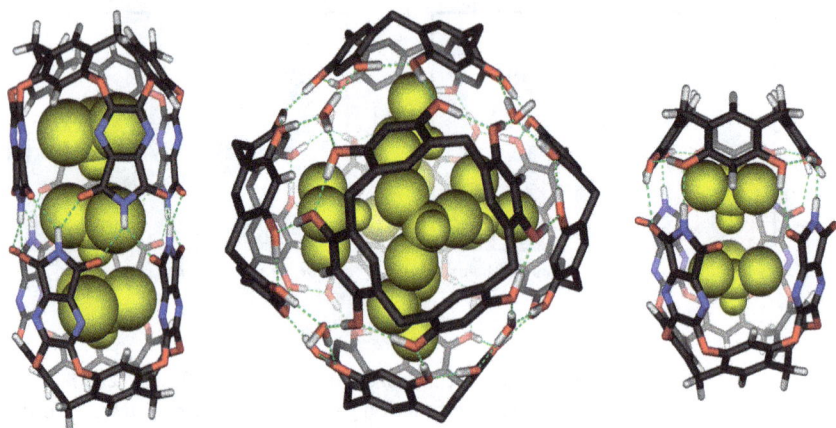

Figure 4.12 *Left*: The cylindrical capsule (Chapter 3) and the resorcinarene hexamer (*center*) spontaneously form a hybrid (*right*). The guests are CHCl$_3$ solvent molecules.

In contrast, Rissanen, Schalley and coworkers showed by mass spectrometry that heterohexamers (hybrids) are formed in the gas phase. Indeed, mixed oligomers of many stoichiometries were observed and characterized.[32] But hybridization with other types of capsules does take place in solution, and FRET techniques were applied to monitor the process in real-time. Specifically, a mixture of the hexameric resorcinarene and the cylindrical capsule (Figure 4.12) formed a heterodimeric species. This occurs in the presence of a guest such as chloroform that can be encapsulated in all the species present.[33] The phenomenon is reversible and the self-sorted state can be regained with the addition of a guest that fits only a specific capsule. For example, benzanilide, which has a high affinity for the cylindrical capsule but is not a guest for the hybrid, drove the re-formation of the self-sorted states.

The self-sorting behavior in the self-assembly of the hexameric resorcin[4]arenes (**1**) and pyrogallol[4]arenes (**2**) capsule phases is difficult to understand. Mattay recently used DOSY techniques to observe a 1:1 complex of resorcinarene with a 2,2,2-trifluoro-1-phenylethanol, a 2:1 complex with 2-butanol and a hexamer with 2-ethyl-octanol.[34] Since no encapsulated guests were apparent in the spectra, these systems appear to be under rapid in–out exchange on the NMR timescale. However, these conditions involve dry solvents, without enough water molecules to complete the seam of hydrogen bonds. In contrast, Aoyama's conditions,[35] in which two-phase extractions from water gave the encapsulated species, showed slow exchange. The unpredictable selectivity of the hexamers for charged species has been brought into focus by recent studies of Philip and Kaifer.[36] They encapsulated the charged cobaltocenium in both hexamers and a series of nitroxides in **1c** (Figure 4.1).[37] The Atwood group was able to load several fluorophores in pyrogallol[4]arenes and study their spectroscopic characteristics, but there the loading was relatively low.[38] At present, there are more examples of guest encapsulation in resorcinarene hexamers than in pyrogallolarene hexamers. Even fewer guests were found to be encapsulated in the hexameric capsule of calixpyridinearene.[9] It should be noted, however, that these observations not only are connected with the affinity of the different guests toward the cavities of **1** and **2** but also reflect the affinity of the solvent used (generally $CDCl_3$) toward the cavity of the respective hexamers.

An elaborate module — a hybrid of the calyxpyrrole (see Chapter 3) and resorcinol — was synthesized by Ballester.[39] While no calixpyrrole-based structure similar to the hexameric resorcinarene was detected, X-ray structures showed hydrogen-bonded capsules in the solid state (Figure 4.13). Subsequent DOSY studies by Cohen[40] established the presence of hexameric aggregates of lipophilic derivatives in $CDCl_3$ solution, but large tertiary amines could not be coaxed inside.

An original and ingenious means of fixing the arrangement of guests within a hexameric host's space was devised by Atwood (Figure 4.14).[41] This involved a hybrid module of one part resorcinol and three parts pyrogallol. On assembly as a hexamer, six (unpaired) hydrogen bond donors are positioned toward the interior of the capsule, where they can pair with donors from guests. The capsule has an internal volume of 860 Å^3, and the figure shows part of the structure of the complex with diethyl ether (there are six encapsulated ethers). Unlike typically disordered guests in resorcinarene and pyrogallolarene hexamers, the guests in the hexamer from the hybrid module

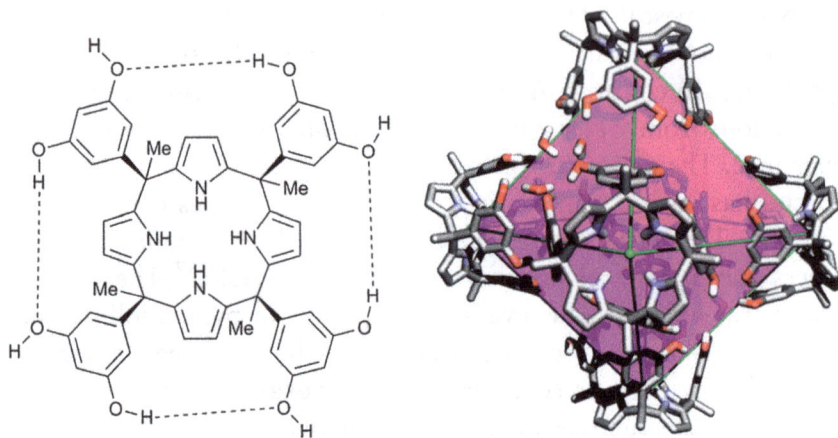

Figure 4.13 *Left*: The formula of a resorcinol fused with a calyxpyrrole. *Right*: X-ray structure of its hexamer in the solid state. [Reprinted with permission from *J Am Chem Soc* **129**: 3820–3821. Copyright 2007, American Chemical Society.]

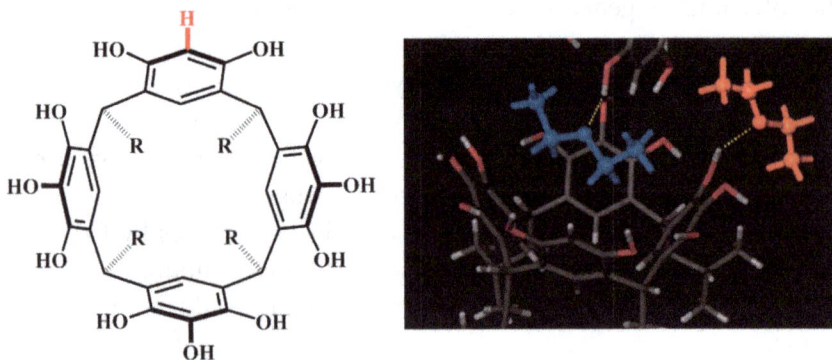

Figure 4.14 *Left*: The formula of a resorcinol covalently bound to three pyrogallols. *Right*: Partial X-ray structure of its hexamer, featuring inwardly directed hydrogen-bonding sites; guests are diethyl ether molecules. [From *Proc Natl Acad Sci USA* **99**: 4837–4841. Copyright 2002, National Academy of Sciences of the United States of America.]

have internal order brought about by specific hydrogen bond donors. The inwardly directed functions augur well for the eventual control of orientations for reactants within the capsule and its use as a reaction flask.

Metallo-supramolecular capsules, which were thoroughly reviewed very recently,[42] have been omitted from this chapter.

References

1. Högberg AGS. (1980) Two stereoisomeric macrocyclic resorcinol-acetaldehyde condensation products. *J Org Chem* **45**, 4498.
2. Kikuchi Y, Tanaka Y, Sutarto S, *et al.* (1992) Highly cooperative binding of alkyl glucopyranosides to the resorcinol cyclic tetramer due to intracomplex guest–guest hydrogen-bonding: solvophobicity/solvophilicity control by an alkyl group of the geometry, stoichiometry, stereoselectivity, and cooperativity. *J Am Chem Soc* **114**, 10302–10306.
3. (a) Sherman JC. (1995) Carceplexes and hemicarceplexes — molecular encapsulation — from hours to forever. *Tetrahedron* **51**, 3395–3422; (b) Timmerman P, Verboom W, Reinhoudt DN. (1996) Resorcinarenes. *Tetrahedron* **52**, 2663–2704.
4. MacGillivray LR, Atwood JL. (1997) A chiral spherical molecular assembly held together by 60 hydrogen bonds. *Nature* **389**, 469–472.
5. Gerkensmeier T, Iwanek W, Agena C, *et al.* (1999) Self-assembly of 2,8,14, 20-tetraisobutyl-5,11,17,23-tetrahydroxyresorc[4]arene. *Eur J Org Chem* 2257–2262.
6. Shivanyuk A, Rissanen K, Kolehmainen E. (2000) Encapsulation of Et_3N^+–H OH_2 in a hydrogen-bonded resorcinarene capsule. *Chem Commun* 1107–1108.
7. Avram L, Cohen Y, Rebek Jr J. (2011) Recent advances in hydrogen-bonded hexameric encapsulation complexes. *Chem Commun* **47**, 5368–5375.
8. Letzel M, Decker B, Rozhenko AB, *et al.* (2004) Encapsulated guest molecules in the dimer of octahydroxypyridine[4]arene. *J Am Chem Soc* **126**, 9669–9674.
9. Gerkensmeier T, Mattay J, Näther C. (2001) A new type of calixarene: octahydroxypyridine[4]arenes. *Chem Eur J* **7**, 465–474.
10. Evan-Salem T, Cohen Y. (2007) Octahydroxypyridine[4]arene self-assembles spontaneously to form hexameric capsules and dimeric aggregates. *Chem Eur J* **13**, 7659–7663.
11. Shivanyuk A, Rebek Jr J. (2001) Reversible encapsulation by self-assembling resorcinarene subunits. *Proc Natl Acad Sci USA* **98**, 7662–7665.
12. Shivanyuk A, Rebek Jr J. (2001) Reversible encapsulation of multiple, neutral guests in hexameric resorcinarene hosts. *Chem Commun* 2424–2425.
13. Avram L, Cohen Y. (2008) Self-assembly of resorcin[4]arene in the presence of small alkylammonium guests in solution. *Org Lett* **10**, 1505–1508.
14. Avram L, Cohen Y. (2002) Spontaneous formation of hexameric resorcinarene capsule in chloroform solution as detected by diffusion NMR. *J Am Chem Soc* **124**, 15148–15149.
15. Evan-Salem T, Baruch I, Avram L *et al.* (2006) Resorcinarenes are hexameric capsules in solution. *Proc Natl Acad Sci USA* **103**, 12296–12300.

16. Avram L, Cohen Y. (2006) Molecules at close range: encapsulated solvent molecules in pyrogallol[4]arene hexameric capsules. *Org Lett* **8**, 219–222.
17. Avram L, Cohen Y. (2006) The role of water molecules in a resorcinarene capsule as probed by NMR diffusion measurements. *Org Lett* **4**, 4365–4368.
18. Ugono O, Holman KT. (2006) An achiral form of the hexameric resorcin[4]arene capsule sustained by hydrogen bonding with alcohols. *Chem Commun* 2144–2146.
19. (a) Slovak S, Avram L, Cohen Y. (2010) Encapsulated or not encapsulated? Mapping alcohol sites in hexameric capsules of resorcin[4]arenes in solution by diffusion NMR spectroscopy. *Angew Chem Int Ed* **49**, 428–1443; (b) Slovak S, Cohen Y. (2012) The effect of alcohol structures on the interaction mode with the hexameric capsule of resorcin[4]arene. *Chem Eur J* **18**, 8515–8520.
20. Slovak S, Cohen Y. (2010) In–out interactions of different guests with the hexameric capsule of resorcin[4]arene. *Supramol Chem* **22**, 803–807.
21. Avram L, Cohen Y. (2004) Self-recognition, structure, stability, and guest affinity of pyrogallol[4]arene and resorcin[4]arene capsules in solution. *J Am Chem Soc* **126**, 11556–11563.
22. Palmer LC, Rebek Jr J. (2005) Hydrocarbon binding inside a hexameric pyrogallolarene capsule. *Org Lett* **7**, 787–789.
23. Yamanaka M, Shivanyuk A, Rebek Jr J. (2004) Kinetics and thermodynamics of a hexameric capsule formation. *J Am Chem Soc* **126**, 2939–2943.
24. Chapin JC, Kvasnica M, Purse BW. (2012) Molecular encapsulation in pyrogallolarene hexamers under nonequilibrium conditions. *J Am Chem Soc* **134**, 15000–15009.
25. Shimizu S, Kiuchi T, Pan N. (2007) A "Teflon-footed" resorcinarene: a hexameric capsule in fluorous solvents and fluorophobic effects on molecular encapsulation. *Angew Chem Int Ed* **46**, 6442–6445.
26. Avram L, Cohen Y. (2003) Discrimination of guests encapsulation in large hexameric molecular capsules in solution: pyrogallol[4]arene versus resorcin[4]arene capsules. *J Am Chem Soc* **125**, 16180–16181.
27. Zhang Q, Tiefenbacher K. (2013) Hexameric resorcinarene capsule is a Brønsted acid: investigation and application to synthesis and catalysis. *J Am Chem Soc* **135**, 16213–16219.
28. Palmer LC, Shivanyuk A, Yamanaka M, Rebek Jr J. (2005) Resorcinarene assemblies as synthetic receptors. *Chem Commun* 857–858.
29. Barrett E, Dale TJ, Rebek Jr J. (2007) Assembly and exchange of resorcinarene capsules monitored by fluorescence resonance energy transfer. *J Am Chem Soc* **129**, 3818–3819.
30. Barrett E, Dale TJ, Rebek Jr J. (2007) Synthesis and assembly of monofunctionalized pyrogallolarene capsules monitored by fluorescence resonance energy transfer. *Chem Commun* 4224–4226.
31. Barrett E, Dale TJ, Rebek Jr J. (2008) Stability, dynamics and selectivity in the assembly of hydrogen-bonded hexameric capsules. *J Am Chem Soc* **130**, 2344–2350.
32. Beyeh NK, Kogej M, Aahman A, *et al.* (2006) Flying capsules: mass spectrometric detection of pyrogallarene and resorinarene hexamers. *Angew Chem Int Ed* **45**, 5214–5218.

33. Ajami D, Schramm MP, Volonterio A, Rebek Jr J. (2007) Assembly of hybrid synthetic capsules. *Angew Chem Int Ed* **46**, 242–244.

34. Schnatwinkel B, Stoll I, Mix A, *et al.* (2008) Monomeric dimeric and hexameric resorcin[4]arene assemblies with alcohols in apolar solvents. *Chem Commun* 3873–3875.

35. Aoyama Y, Tanaka Y, Toi H, Ogoshi H. (1988) Polar host–guest interaction. binding of nonionic polar compounds with a resorcinol-aldehyde cyclooligomer as a lipophilic polar host. *J Am Chem Soc* **110**, 634–635.

36. (a) Philip I, Kaifer AE. (2002) Electrochemically driven formation of a molecular capsule around the ferrocenium ion. *J Am Chem Soc* **124**, 12678–12679; (b) Philip I, Kaifer AE. (2005) Noncovalent encapsulation of cobaltocenium inside resorcinarene molecular capsules. *J Org Chem* **70**, 1558–1564.

37. Mileo E, Yi S, Bhattacharya P, Kaifer AE. (2009) Probing the inner space of resorcinarene molecular capsules with nitroxide guests. *Angew Chem Int Ed* **48**, 5337–5340.

38. (a) Dalgarno SJ, Tucker SA, Bassil DB, Atwood JL. (2005) Fluorescent guest molecules report ordered inner phase of host capsules in solution. *Science* **309**, 2037–2039; (b) Bassil DB, Dalgarno SJ, Cave GWV, *et al.* (2007) Spectroscopic investigations of ADMA encapsulated in pyrogallol[4]arene nanocapsules. *J Phys Chem B* **111**, 9088–9092.

39. Gil-Ramirez G, Benet-Buchholz J, Escudero-Adan EC, Ballester P. (2007) Solid-state self-assembly of a calix[4]pyrrole–resorcinarene hybrid into a hexameric cage. *J Am Chem Soc* **129**, 3820–3821.

40. Slovak S, Evan-Salem T, Cohen Y. (2010) Self-assembly of a hexameric aggregate of a lipophilic calix[4]pyrrole–resorcinarene hybrid in solution: a diffusion NMR study. *Org Lett* **12**, 4864–4867.

41. Atwood JL, Barbour LJ, Jerga A. (2002) Organization of the interior of molecular capsules by hydrogen bonding. *Proc Natl Acad Sci USA* **99**, 4837–4841.

42. See, e.g.: Dalgarno SJ, Power NP, Atwood JL. (2008) Metallo-supramolecular capsules. *Coord Chem Rev* **252**, 825–841; (b) Jin P, Dalgarno SJ, Atwood JL. (2010) Mixed metal-organic nanocapsules. *Coord Chem Rev* **254**, 1760–1768.

CHAPTER 5

Stereochemistry of Confined Molecules

The translational consequence of molecular encapsulation is fairly obvious: the guest molecule's entropy is reduced as a result of its confinement in a small space. But what about its other motions — spinning, tumbling, internal rotations and vibrations? While there is never a fit between host and guest that is snug enough to effectively alter vibrations, we will see that other internal motions such as sigma bond rotations can be profoundly affected and result in otherwise unknown contorted conformations.

Carceroisomerism

Encapsulation of a guest molecule that has low symmetry will immediately have consequences for the spectroscopic characteristics of the host capsule. If the molecule is free to tumble within the space, the spectroscopic consequences might be slight. This is particularly true if the capsule is highly symmetrical — and most are, as a consequence of their assembly from identical modules. Reinhoudt and coworkers made a covalent capsule of reduced symmetry consisting of two different hemispheres.[1] They showed that the binding of a molecule led to two isomers when the molecule was not free to tumble inside. They called this new feature of limited space "carceroisomerism." The strength of the covalent bonds that held the capsule together was responsible for creating and maintaining this isomerism. However, not all covalent capsules have the

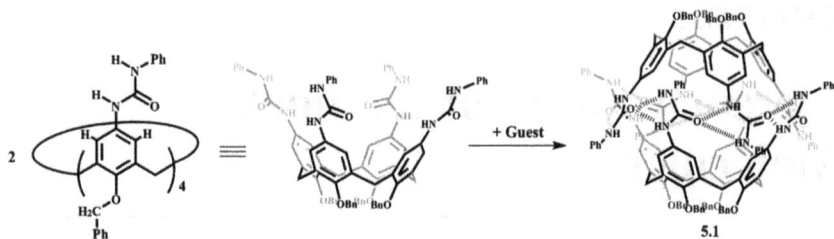

Figure 5.1   The calixarene cavitand and its capsule **5.1**. Hydrogen bonding of the head-to-tail urea functions holds the hemispheres together.

capacity for such isomerism. For example, the cryptophane is simply too small.[2]

Our first encounter with this type of isomerism was in the capsules derived from the calixarenes (Figure 5.1). These capsules will also be encountered in the sections on chirality (Chapter 6) but here we emphasize the confinement-induced stereochemical features. Ron Castellano observed that with guests such as certain terpenes (Figure 5.2) that made for relatively snug fits inside a symmetrical capsule, two separate complexes were formed. For example, tricyclene **5.5** gave two complexes of different populations; the complexes have different energies, and a large energetic barrier separates them.[3] There are preferred orientations of guests inside the symmetrical hosts, as shown by the spectra of Figure 5.2, but a more general statement is simply that the motion of the guest inside the host was not free.

The heterodimeric calixarenes of ureas with sulfonylureas described in Chapter 6 provide a more complicated situation. These studies, done by Ron Castellano and Byeang Hyeam Kim, showed that the heterodimers exist as a pair of enantiomers and a chiral guest such as nopinone **5.2** will naturally lead to diastereomeric complexes.

Social Isomerism

Different Co-guests

Restricted motion of *two* molecules in a small space also gave rise to new forms of isomerism. The shaped space of the cylindrical

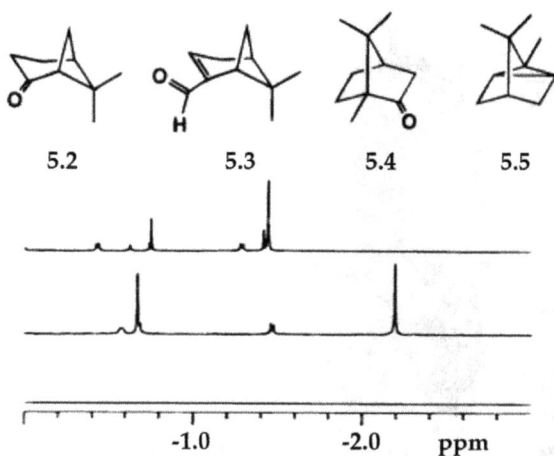

Figure 5.2   *Top*: Small molecule guests of the calixarene capsule. A single complex is observed by NMR with nopinone **5.2**; (*top trace*) but two complexes of unequal energy are observed with ticyclic **5.5**; (*bottom trace*). [Reprinted with permission from *J Am Chem Soc* **119**, 12671–12672. Copyright 1997, American Chemical Society.]

capsule is ideal for such studies, and we have exploited these as follows. Alex Shivanyuk and Alessandro Scarso added the capsule to a mixture of chloroform and *p*-ethyltoluene and observed two encapsulated assemblies **5.6** and **5.7** in the NMR spectra (Figure 5.3). The peaks are sharp and widely separated, which speaks for a large energetic barrier between the two isomeric arrangements, and their interconversion is slow on the NMR timescale. The chloroform is too large to slip past the co-guest, which would be one way of interconverting these isomers while they are inside the capsule. The *p*-ethyltoluene (or practically any *p*-disubstituted benzene) is too long to tumble inside the capsule; this would also interconvert the isomers. We call these social isomers because it takes two molecules inside to expose this type of spatial arrangement. These are constitutional isomers, and in some ways they are also diastereomers, but in classical organic chemistry diastereomers are related through their covalent "connectedness." In this case, they are related not by covalent but by mechanical connectedness. The walls of the capsule fix them in space and maintain their relationship to each other.

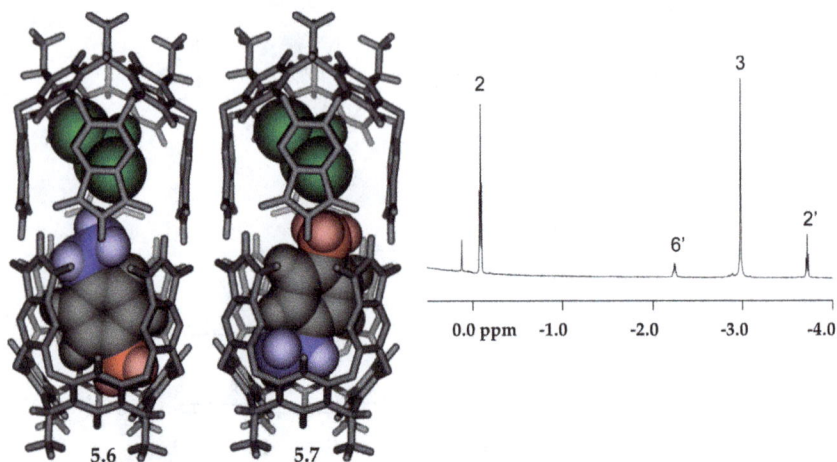

Figure 5.3 Social isomers of $CHCl_3$ and *p*-ethyltoluene do not interconvert while inside the capsule. The co-guests cannot slip past each other and the *p*-disubstituted aromatic is too long to tumble freely inside: two sets of signals appear in the NMR spectrum. [Reprinted with permission from *J Am Chem Soc* **125**, 13981–13983. Copyright 2003, American Chemical Society.]

At first glance, it would seem that the two isomers should have roughly equal energies, but they do not. The isomers appear in unequal concentrations, which can be used to calculate their difference in free energies, $\Delta G°$'s. The origins of these differences in energies are, unfortunately, inseparable components because of the capsule itself. First, and inherently, a methyl group fits the tapered end of the cavity better than, say, an ethyl group, and this causes a bias in the orientation of *p*-ethyltoluene. Second, there are possible interactions *between* the two guests, which will also bias the arrangements.

We arranged to encapsulate certain pairs of social isomers in order to probe whether a pattern of interactions could be discerned between the two guests in a small space. In solution, the random flurry of collisions and rapid exchange of partners make such directional interactions impossible to evaluate, but we regard these as single solvent–solute interactions. The distinction between the solvent and the solute is arbitrary. Alex Shivanyuk and Alessandro Scarso studied the interaction of benzene with *p*-ethyltoluene (Figure 5.4).[4]

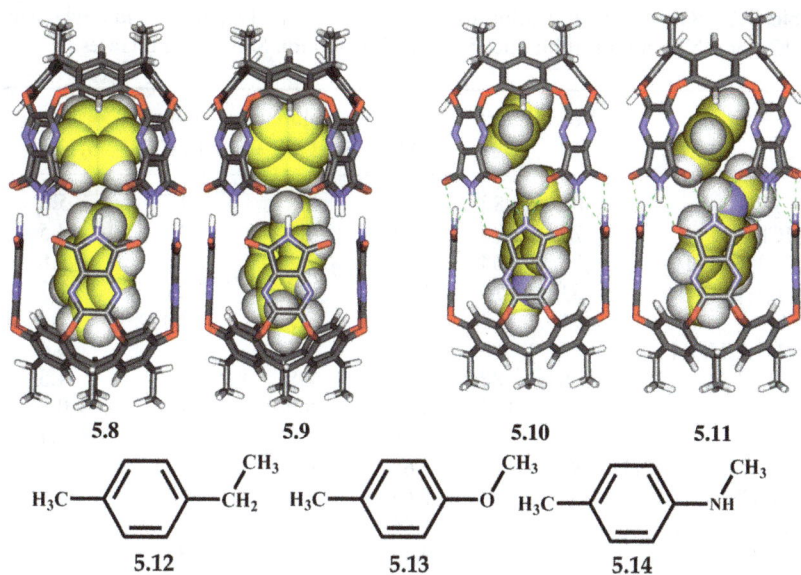

Figure 5.4 Social isomers of benzene with (*left*) *p*-ethyltoluene **5.12** and (*right*) N-methyl-*p*-toluidine **5.14**.

There is a sixfold preference (**5.8** 86% to **5.9** 14%) corresponding to a lower energy of ~1 kcal/mol for the social isomer that has the ethyl group near the benzene versus the ethyl in the tapered end of the cavitand. However, merely replacing the CH_2 with an NH rewrites these preferences; now the NH-methyl prefers the cavitand (**5.10** 60%) to the contact with the benzene co-guest (**5.11** 40%). The geometry of the space requires that two aromatic guests encounter each other largely edge-to-edge. A "coaxial" CH_3 group fits nicely into the cavitand end and one could expect the more polar $NH-CH_3$ to drift toward the seam of hydrogen bonds. These factors would favor **5.11**. But the $CH-\pi$ interactions of the $NH-CH_3$ are better with the resorcinarene than with the edge of the benzene co-guest. When the $NH-CH_3$ is changed to the $O-CH_3$ of an anisole, the preferences revert to the original (CH_2-CH_3) case. The three *p*-substituted toluenes (OMe, NHMe and Et) as solutes were coencapsulated with a panel of common solvents, and the social isomer ratios are shown in Table 5.1. What are the origins of these preferred arrangements in space? There

Table 5.1 Social Isomer Ratios of p-Ethyltoluene (5.12), with 4-methylanisole (5.13), and N-methyl-p-toluidine (5.14) with Common Solvents and Gases.

Solvent	Vol.[a] $Å^3$	Surf. $Å^2$	Dipole D^b	Solute 5.12 $K^c =$	Solute 5.13 $K^c =$	Solute 5.14 $K^c =$
CH_2Cl_2	60	82	1.60	2.5	4.7	5.6
$CHCl_3$	75	98	1.01	4.4	3.1	3.5
$(CH_3)_2CHCl$	76	104	2.17	2.2	2.7	1.8
$(CH_3)_2CCl_2$	91	119	2.27	4.6	2.8	1.7
CCl_4	91	113	0	5.7	2.1	1.4
$(CH_3)_2CHOH$	66	96	1.68	4.1	4.5	4.4
$(CH_3)_2CO$	60	88	2.88	2.8	5.6	4.3
C_6H_6	77	103	0	6.0	1.2	0.7
C_6H_5N	72	98	2.19	4.3	4.3	3.1
$C_2H_5(CH_3)CHOH$	82	116	1.64	4.5	3.2	2.4
$C_2H_5CH_2OH$	66	96	1.68	3.5	7.1	8.9
$C_3H_7CH_2OH$	81	118	1.66	4.0	4.6	13.5
$C_2H_5CH(CH_3)_2$	91	124	0.13	5.5	2.5	1.6
n-C_5H_{12}	90	127	0	2.4	1.9	1.7
c-C_6H_{12}	97	127	0	10	1.2	0.95
$CH_3CH=CH_2$	54	78	0.37	2.9	3.2	4.6
CH_3CH_3	42	66	0	2.8	4.4	4.1

[a]Volumes and surfaces were minimized with the program Hyperchem 7.0, Hypercube, Inc., 2002, at semi-empirical PM3 level and calculated with WebLab Viewer Pro 4.0 by Molecular Simulations, Inc. [b]Dipole moments are from the *Handbook of Chemistry and Physics*. [c]The social isomeric ratios are subject to uncertainties of ±10%, due to integration errors.

is no secure answer, only factors that appear *ad hoc* rather than generalizable explanations. Although the social isomer system has sub-kilocalorie sensitivity and, like Wilcox's torsion balance,[5] exists in only two states, the interpretations are much less obvious.

Identical Co-guests

The requirement for p-disubstitution was established with shorter guests such as toluene. Two molecules of toluene are encapsulated, but a single resonance is observed at about −0.6 ppm for the methyl group. This signal represents the weighted average for the magnetic environments of all the social isomers present, but these interconvert

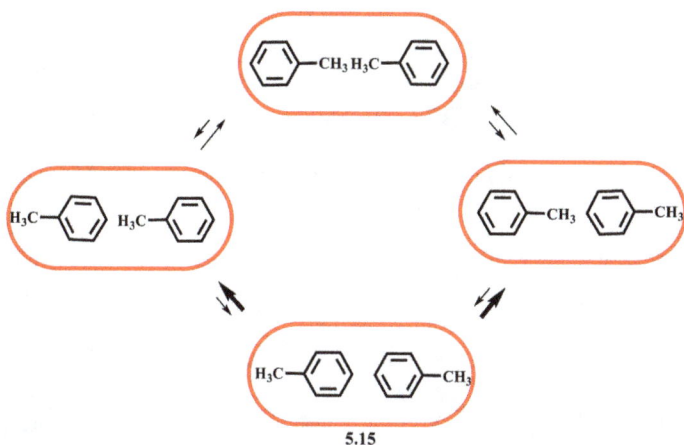

Figure 5.5 Tumbling of encapsulated toluene in the capsule is rapid on the NMR timescale and an averaged signal at -0.6 ppm is observed. The isomer ratios calculated by NICS indicate that 60–70% of the CH_3 groups are near the middle of the capsule. [Reprinted with permission from *Chem Eur J* **19**(50), 17092–17096. Copyright 2013, Wiley-VCH, Weinheim.]

rapidly by tumbling as shown in **5.15** (Figure 5.5). Attempts to freeze out the tumbling have failed, and the analysis of the spectra led to an estimate that about 30–40% of the CH_3 groups are near the capsule ends at any given time. With a benzene guest bearing an ethyl or longer group, the tumbling cannot occur while in this capsule.

But even the dynamic systems that tumble can be interpreted with some accuracy. Dariush Ajami unraveled the finer details of the social isomers of encapsulated picolines with the computational support of our collaborators, Giannoula Theodorakopoulos and Ioannis Petsalakis.[6] Earlier work by Fabio Tucci and Dmitry Rudkevich had established that two molecules of any combination of picolines were encapsulated,[7] and led to NMR spectra interpretations that we based on a static model. With the discovery of social isomers and the application of nucleus-independent chemical shifts (NICS) calculations devised by Schleyer,[8] a dynamic interpretation was developed and gave the results of Figure 5.6.

The picolines share several features: their dipole moments are comparable ($\alpha = 2.1$, $\beta = 2.5$, and $\gamma = 2.7$ D),[9] they have

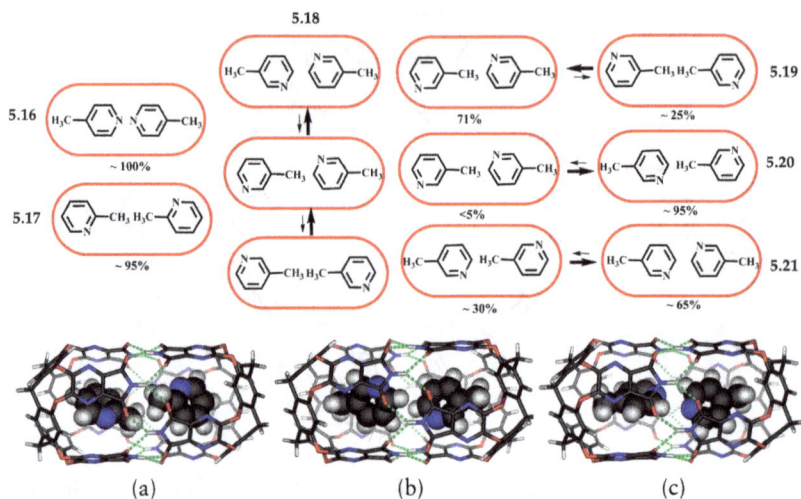

Figure 5.6 Orientation of homodimeric picolines in the capsule. The α- and γ-picolines show strong preferences but the β-picoline is mobile. Heterodimers **5.19–5.21** can be analyzed through NICS calculations.

Views along the long axis of the capsule

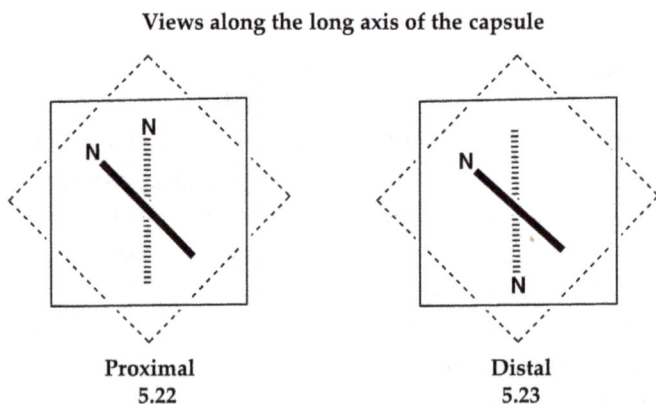

Proximal
5.22

Distal
5.23

Figure 5.7 Cartoons of coencapsulated α- or β-picolines. The two nitrogen arrangements can interconvert by a 90° rotation of one guest on the long axis of the capsule.

attractions to the seam of the hydrogen bonds near the capsule's center guest's polar areas, and their (polarized) methyl groups can participate in CH–π attractions with the resorcinarenes. Both γ- and α-picolines evince a single, dominant social isomer (Figure 5.7), and yet they are quite different, as depicted in **5.16** and **5.17**, respectively. Calculations suggest that hydrogen bonds exist between the coencapsulated γ-picolines. For the α-picolines, the nitrogens appear to be directed toward the carbonyls of the capsule's walls. Unaccountably, β-picoline does not find a most comfortable arrangement in the capsule: on average, about 65% of its N-atoms are directed "inwardly," while 35% are directed "outwardly," as indicated in **5.18**.

Heterodimeric Picolines

While it is not our intent to describe the detailed interactions of heterodimeric complexes (for example, α- and β-picolines in the same capsule), we feel obligated to summarize the outcomes in Figure 5.6 to correct the egregious error of a similar figure in the original published account. The simple rule seems to be that *the N-atoms prefer to be near the center of the capsule*. This is rigorously true for the γ-isomer, which prefers (>90%) an inwardly directed N and outwardly directed methyl, regardless of the co-guest. For α-picoline again 90% of the N (and methyls) are inwardly directed. For β-picoline the inward N is 60–70% of the arrangements, depending on the co-guest.

Further Rotational Isomerism

The very shape of the space inside the capsule — two square prisms rotated 45° to each other — gives rise to additional possibilities for isomerism. The two resorcinarene ends are tapered but the imide walls of the capsule define a box and any cross-section shows a square shape. Simple aromatics fit best when positioned diagonally across this square cross section as in Figure 5.7. Aromatic guests in the two halves cannot be coplanar, as described for the twisted conformation of stilbene in Chapter 3. Consider now the α,β-picoline homodimers or the α,β-picoline heterodimers inside the capsule. In their ground states the N–atoms can be proximal **5.22** or distal **5.23** to one another, as shown in the cartoons. These are social isomers. When one guest spins 90° on the long axis of the capsule, the isomers interconvert. The breathing motions of the walls distort the hydrogen bonds and permit the rapid rotation and, at the present time, we have been unable to freeze it out. With narrower spaces (or thicker guests) the energetic barrier to rotation could be raised to reveal these additional social isomers and increase the information content of the complexes as discussed below.

Constellational Isomers

Solvent Arrays

The third new stereochemical relationship exposed in the cylindrical space was uncovered by Alex Shivanyuk.[10] He found that the capsule took up three small solvents, such as $CHCl_3$. He dissolved the capsule in $CHCl_3$, then evaporated the solvent and heated the residue under reduced pressure overnight. When he redissolved what remained in MeOH, exactly three $CHCl_3$ molecules were released from each capsule.

But the three $CHCl_3$ guest showed only two NMR signals: one shifted modestly upfield for the centrally located $CHCl_3$, and another signal of twice the intensity that was shifted more strongly upfield for the $CHCl_3$'s at either end of the capsule. The signals were sharp, indicating that the guest positions did not exchange on the NMR timescale and a large energetic barrier existed for motion between the two environments. The $CHCl_3$'s are simply too large to move past each other while inside the cylindrical space. Entirely analogous results were obtained for isopropyl chloride as a guest — and why not? The size and shape of $CHCl_3$ and isopropyl chloride are almost identical. The only difference was that with isopropyl chloride guests the NMR signals were in an easily interpreted region of the spectrum (Figure 5.8, trace a).

With the longer and narrower 1,2-dichloroethane, similar results were obtained, but guest signals were slightly broadened, and the signal for the centrally located guest was obscured by the signals of the alkyl groups on the feet of the capsule. However, a sample of the capsule prepared from perdeuterated dodecanal revealed the position of the centrally located guest signal through a 1D GOESY

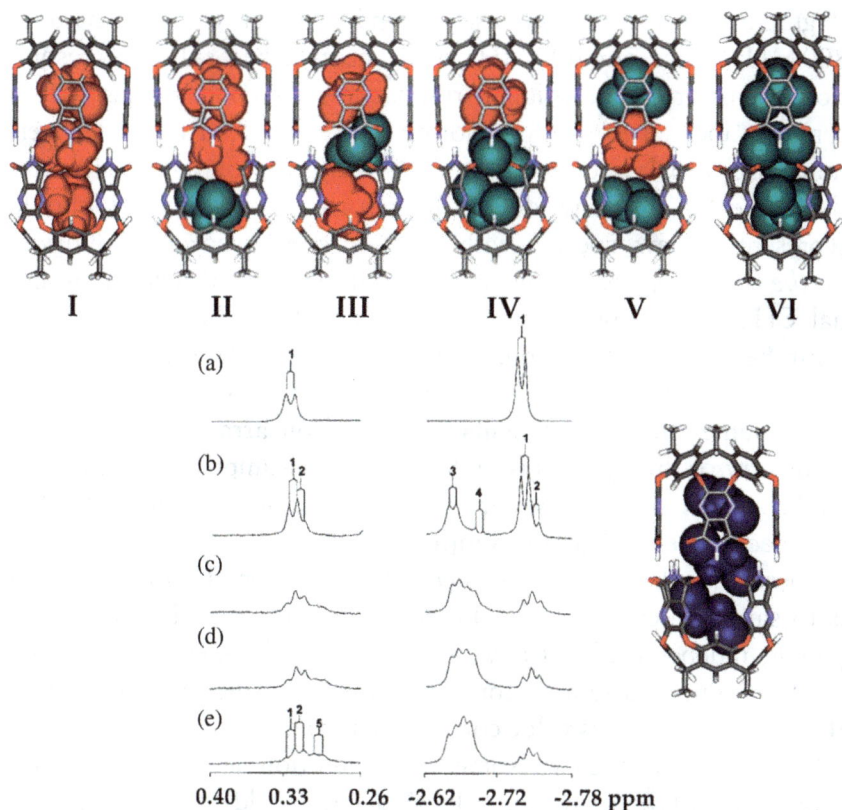

Figure 5.8 *Top*: Possible arrangements for coencapsulated CHCl₃ (green) and isopropyl chloride (red). *Bottom*: Upfield regions of the 800 MHz ¹H NMR spectra during titration of an isopropyl-chloride-filled capsule **I** with CHCl₃. Only the isopropyl guest appears in the regions shown. Arrays **II** and **III** (and **IV** and **V**) are isomeric constellations. [Reprinted with permission from *Angew Chem Int Ed Eng* **42**, 684–686. Copyright 2002, Wiley-VCH Weinheim.]

experiment.[11] This experiment also showed that *occasionally* these thinner and more flexible guests could snake past each other while still in the capsule. Even so, it was clear that the four possible arrangements of CHCl₃ and isopropyl chloride as coguests should be stable on the NMR timescale, and we set out to find and characterize them.

When a small amount of CHCl₃ is added to a capsule filled with isopropyl chloride, the NMR spectrum changes dramatically

(Figure 5.8, trace b). We used a spectrometer that operated at 800 MHz, since our 600 MHz instrument lacked sufficient resolution. Apart from the original resonances, a second species with a centrally located isopropyl chloride is present. And three new terminal isopropyl resonances appear. Only one of these integrates the same as the new downfield resonance, so this set should represent a new species with *two different isopropyl chlorides in the same capsule*. This isomer can only be **II** with a terminal CHCl₃. The two additional new signals in the upfield region must be **III** and **IV** — but which is which? Mere statistics suggest that the signal with the (much) larger integration is **III**, but this assumes that no particular combination arrangement or stoichiometry of these guests is preferred. The assumption is reasonable as the smaller signal grows at the expense of others as more CHCl₃ is added (trace c). There is another relationship that exists between isomeric capsules **II** and **III** — their ratios must remain constant. That is to say, whatever energy differences exist between **II** and **III** will persist, independently of any other assemblies present.[12] Their relative amounts did remain constant throughout the titration, with a **II:III** ratio of 0.4. This value corresponds to a difference in energies of ~0.5 kcal/mol for the two isomers. The addition of more CHCl₃ also gives a third type of centrally located (downfield) isopropyl group, namely the isomer **V**, with an isopropyl chloride sandwiched between two CHCl₃'s. The presence of all isopropyl-containing capsules in trace c is evidence that the isomers are more or less of the same energy and statistically determined. The isopropyl chloride does seem to prefer the ends of the capsule (compare **IV** and **V**). A corollary is that a CHCl₃-filled capsule **VI** is also present in the sample of trace c (but those guests' resonances do not appear in this region of the spectrum). The constellational isomers of CHCl₃ with dichloroethane and those of isopropyl chloride with dichloroethane were also prepared and characterized by NMR. The three capsules with one each of these guests remain to be identified.

What is the relationship between the isomers **II** and **III** or the isomers **IV** and **V**? We named these new relationships "constellational" isomers. As was the case of social isomers, they are diastereomeric arrangements with their specific connectedness enforced by the mechanical barriers of the capsule's walls. With two possible guests — $CHCl_3$ and isopropyl chloride — six arrangements are possible, including those capsules with three of the same guests inside. These arrays represent information, and they might be used for data storage: if they can be "written" at will, maintained for longer times and "read out" rapidly, six bits of information could be stored. With three possible guests, 18 arrangements can be packed in the space of a few cubic nanometers — truly nanoscale data storage. We have encountered one of these in Chapter 3, where a PF_6^- ion is encapsulated between a molecule of $CHCl_3$ and isopropyl chloride. But since the "writing" and reading depend on molecular (rather than electronic) motion, a notional computer made with such components would operate at the speed of sound rather than the speed of light. This represents a serious disadvantage.

Chiral Assemblies

The encapsulation of chiral molecules brings with it the possibility of polarimetric readout of capsule content. For example, three molecules of propylene sulfide are encapsulated, and the guests (63 $Å^3$ each) fill about 45% of the space in the capsule. This is a low value of the packing coefficient for liquids, but a high one for gases. Since there are an odd number of guests inside, *every capsule is chiral*.[13] However, each contellational isomer pair has the same specific rotation. To identify them all, Masamichi Yamanaka prepared an optically active sample of the sulfide.[14]

Two sets of methyl groups were observed −2.59 ppm and 0.23 ppm, in a 2:1 ratio (Figure 5.9(a)). These represent encapsulated guests at the ends of the capsule and the center, respectively. A 2D NOESY spectrum showed no cross-peaks between upfield and downfield signals, but cross-peaks based on chemical exchange between encapsulated propylene sulfides and propylene sulfide in bulk solution were observed.

The NMR spectrum of the *racemic* guest at 243 K showed four well-resolved doublets around −2.7 ppm. When all isomers are present, four sets of methyl protons can be observed around −2.7 ppm

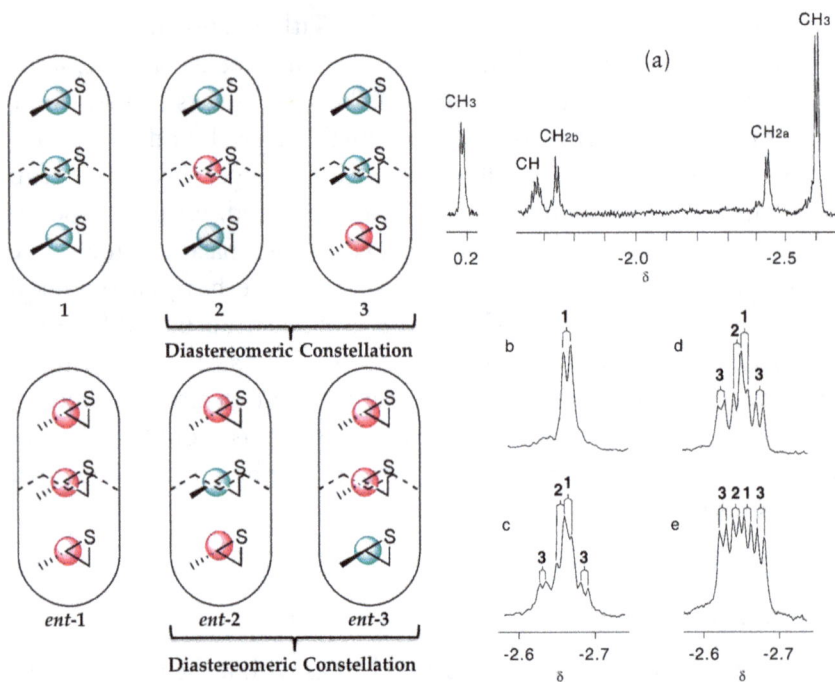

Figure 5.9 *Left*: Possible constellations of propylene sulfide. Upfield regions of the 600 MHz ^{1}H NMR. (a) (R)-propylene sulfide (88% ee) at 300 K, (b) (R)-propylene sulfide (88% ee) at 243 K, (c) (R)-propylene sulfide (60% ee) at 243 K, (d) (R)-propylene sulfide (30% ee) at 243 K and (e) (±)-propylene sulfide at 243 K. [Reprinted in part from *Chem Commun* 1690–1691 (2004) with the permission of The Royal Society of Chemistry.]

(Figure 5.9e). These comprise one set from isomer **1**, one set from isomer **2** and two sets from isomer **3**. Assignment of those signals was enabled by experiments using varying enantiomeric excesses (ee's). In the case of 60% ee, the signals were sufficiently resolved to assign all isomers (Figure 5.9c). Further reduction of the ee of the guest solution used resulted in an increase in the relative intensities of isomers **2** and **3** with their mirror images (Figure 5.9d). These spectra are characteristic of the ee's of the guests, and it is possible to estimate their optical purity by NMR without any other information. We dwell on this process because it could be applied to other small chiral molecules.

Pyrogallolarenes

Recent work by Cohen has revealed that small solvent molecules — CH_2Cl_2, $CHBr_3$ and CH_3I, but not $CHCl_3$ — can occupy distinct sites within self-assembled pyrogallolarene hexamers (Chapter 2). Even though the spaces on these hydrogen-bonded capsules are pseudospherical, there appear to be separate peripheral and central sites and guests are in slow exchange on the NMR timescale between the sites.[15] While the details are not yet elucidated, the possibilities for additional types of isomerism are evident, and suggest that there is much to be discovered for small guests in these confined spaces.

References

1. Timmerman P, Verboom W, van Veggel FCJM, *et al.* (1994) A novel type of stereoisomerism in calix[4]arene-based carceplexes. *Angew Chem Int Ed Eng* **33**, 2345–2348.
2. Canceill J, Cesario M, Collet A, *et al.* (1985) A new bis-cyclotribenzyl cavitand capable of selective inclusion of neutral molecules in solution — crystal-structure of its CH_2Cl_2 cavitate. *J Chem Soc Chem Commum* **6**, 361–363.
3. Castellano RK, Kim BH, Rebek Jr J. (1997) Chiral capsules: asymmetric binding in calixarene-based dimers. *J Am Chem Soc* **119**, 12671–12672.
4. Scarso A, Shivanyuk A, Rebek Jr J. (2003) Individual solvent/solute interactions through social isomers. *J Am Chem Soc* **125**, 13981–13983.
5. Paliwal S, Geib S, Wilcox CS. (1994) Molecular torsion balance for weak molecular recognition forces. effects of "tilted-T" edge-to-face aromatic interactions on conformational selection and solid-state structure. *J Am Chem Soc* **116**, 4497–4498.
6. Ajami D, Theodorakopoulos G, Petsalakis ID, Rebek Jr J. (2013) Interactions and arrangements of picolines in a small space. *Chem Eur J* **19**, 17092–17096.
7. Tucci FC, Rudkevich DM, Rebek Jr J. (1999) Stereochemical relationships between encapsulated molecules. *J Am Chem Soc* **121**, 4928–4929.
8. Schleyer PvR, Maerker C, Dransfeld A, *et al.* (1996) Nucleus-independent chemical shifts: a simple and efficient aromaticity probe. *J Am Chem Soc* **118** 6317–6318.
9. Gramstad T. (1993) Studies of hydrogen bonding. Part XXVI. Dipole moments of pyridines, quinolines, and acridine, and of their hydrogen-bonded complexes with phenol. *Acta Chem Scand* **47**, 985–989.
10. Shivanyuk A, Rebek Jr J. (2003) Isomeric constellations of encapsulation complexes store information on the nanometer scale. *Angew Chem Int Ed* **42**, 684–686.

11. Shivanyuk A, Rebek Jr J. (2002) The inner solvation of a cylindrical capsule. *Chem Commun* 2326–2327.
12. Yamanaka M, Rebek Jr J. (2004) Stereochemistry in self-assembled encapsulation complexes: constellational isomerism. *Proc Natl Acad Sci USA* **101**, 2669–2672.
13. Yamanaka M, Rebek Jr J. (2004) Constellational diastereomers in encapsulation complexes. *Chem Commun* 1690–1691.
14. Hauptman E, Fagan PJ, Marshall W. (1999) Synthesis of novel (P, S) ligands based on chiral nonracemic episulfides. Use in asymmetric hydrogenation. *Organometallics* **18**, 2061–2073.
15. Guralnik V, Avram L, Cohen Y, Unique organization of solvent molecules within the hexameric capsules of pyrogallol[4]arene in solution. *Org Lett* **16**, 5592–5595.

CHAPTER 6

Chiral Capsules

There are many approaches to the creation of chiral capsules and we have tried them all. Generally speaking, the chirality can be outside the capsule, in its lining or — most desirably — inside the capsule's space. We take up these locations here for different capsules but warn the reader that none proved deeply satisfying: hydrogen bonds are elastic and the shapes of the space they enclose are easily distorted.

Chiral Outside

The simplest case of a chiral capsule with an achiral space involves dimerization with modules of reduced symmetry. This results in a chiral — but of course, racemic — capsule, such as the tennis ball module that featured two different glycolurils (Chapter 1). That chiral capsule was useful for studying the rate of dissociation and recombination of capsule halves (racemization), but since the asymmetry was due to features on the outer surface of the capsule, it did not result in a chiral inner space. We will later relate the consequences for the guest's *magnetic* environment, but we had no chiral probe small enough to fit inside the tennis ball, anyway. The smallest covalent capsule, a cryptophane can accommodate something the size of bromochlorofluoromethane[1] but we had our sights on larger targets. With access to the larger softball that could accommodate guests such as camphor derivatives, it was possible to test the effects of the outer surface on the inner space. Yuji Tokunaga prepared the appropriate softball[2] (Figure 6.1) and examined its uptake of chiral guests in p-xylene-d_{10} (itself a poor guest). In all cases diastereomeric complexes were formed in equal concentrations, confirming that the asymmetry of the outer

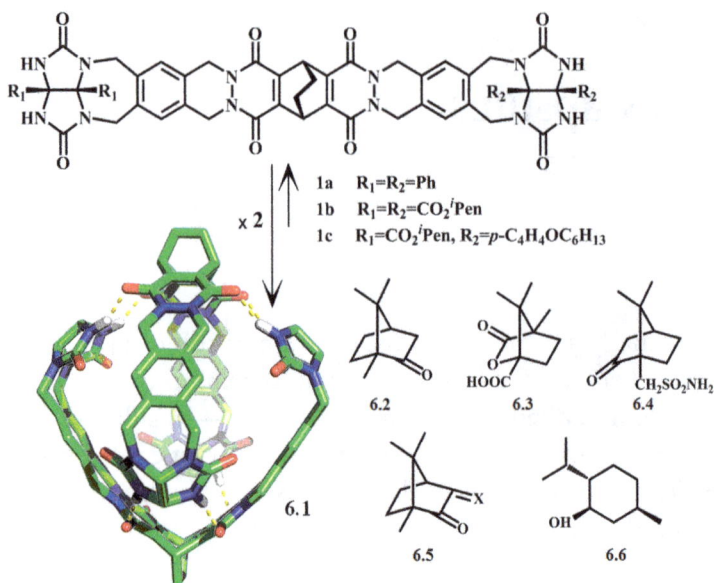

Figure 6.1   A softball **6.1** with a chiral outer surface. Typical chiral guests such as camphor **6.2** and other terpenes **6.3–6.2** showed no enantioselective encapsulation.

surface of the capsule does not extend into the shape of the cavity. The guest is affected in a magnetic sense, since separate signals for the diastereomers are observed, but the asymmetry is not transmitted in a way that distorts the inner space in a steric sense since the complexes have equal energies.

Chiral Lining

The calixarene capsules as the octa-ureas prepared from aromatic iso-cyanates (Chapter 2) feature S_8 symmetry and are achiral. Ron Castellano and Byeang Hyean Kim used optically active isocyanates derived from α-phenylethylamine and amino acid esters in an attempt to make the enantiopure capsules shown in Figure 6.2.[3] The branching at the asymmetric centers destabilized the homodimeric capsules but heterodimeric complexes **6.7** could be formed with the simpler aryl ureas. The asymmetric centers influence the clockwise/counterclockwise

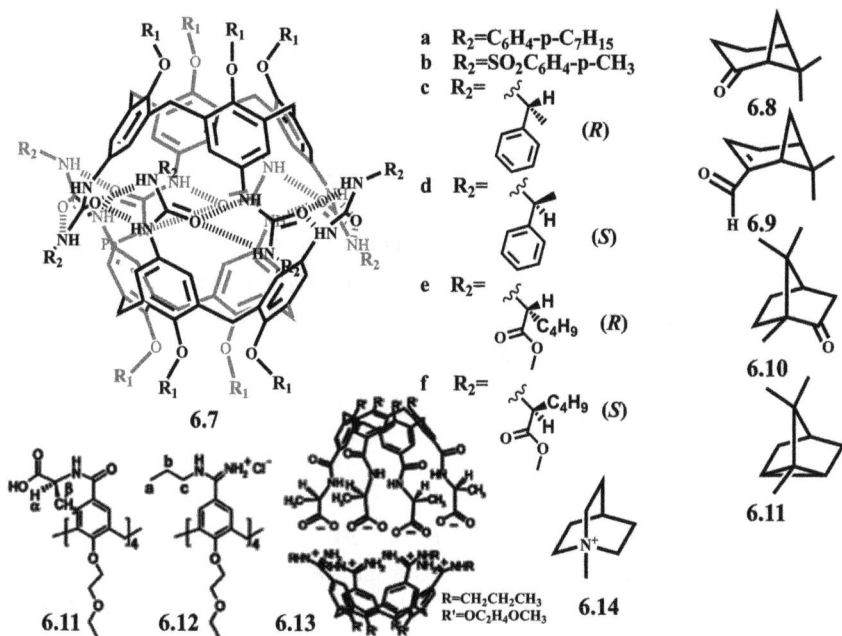

Figure 6.2 *Top*: Capsules of the calixarene ureas and typical terpene guests. *Bottom*: A water-soluble capsule held together by salt bridges **6.13** binds N-methyl quinuclidinium ion **6.14**.

sense of the head-to-tail ureas, i.e. the lining of the capsule is chiral and enantiopure. The NMR spectra were profoundly affected, since the magnetic environment of both the capsule and guests such as terpenes was influenced by the many asymmetric centers. However, the enantioselectivity of the capsule itself was very modest — at best, a 15% excess of one diastereomeric complex could be obtained with typical terpene guests. In short, the steric features remained on the outside and lining of the capsule, but did not intrude to affect the space inside. The tight fit of these guests in capsules did give rise to a stereochemical feature known as "carcerisomerism,"[4] where different orientations of the guest are held in space and interconvert slowly on the NMR timescale.

There are other hydrogen-bonded capsules which show host–guest recognition in that most unlikely solvent — water. One example, due to Reinhoudt,[5] uses the calixarene platform and involves ionic interactions between carboxylates on one hemisphere **6.11** and amidiniums **6.12** on the other (Figure 6.2). These groups are known to have hydrogen-bonding complementarity as well, but the propyl groups of the amidiniums are directed toward (and observed inside) the hemisphere of the opposite carboxylates. This arrangement must hinder hydrogen bonding, especially the two-point arrangement that is ideal between these donors and acceptors. Nonetheless, encapsulation of N-methyl quinuclidinium in **6.14** was observed **6.13** under conditions of rapid exchange in solution by NMR, and in the gas phase by ESI mass spectrometry.

Chiral Twist

A spectacularly complex capsule, based on the signature double rosette motif of Reinhoudt, is held together by 36 hydrogen bonds (Figure 6.3). The rosettes **6.15**, which can be obtained in chiral form, have sizable cavities and bind anthracene **6.16** and its derivatives such as alizirin as trimers in organic solvents ($CDCl_3$).[6] The asymmetric centers are on the capsule's **6.17** periphery, where they transmit chiral information by inducing a twist between the top and bottom rosettes.

Chiral Inside

Racemates

The desired goal was to create a chiral *space* and this was achieved with the softball. Jose Rivera and Tomas Martin incorporated different spacers **6.17** and **6.18** between the central bicyclic system and the glycoluril ends (Figure 6.4). This desymmetrizes the shape of the space inside. As was the case in the synthesis of desymmetrized tennis balls, the capsules are chiral but racemic. The modules have curvature that distinguishes "front" from "back" and the different spacers distinguish "left" from "right" in the depiction shown, but they still have a plane of symmetry and so the "top" and "bottom" are the same.[7] The shape of the resulting space is a somewhat distorted sphere with

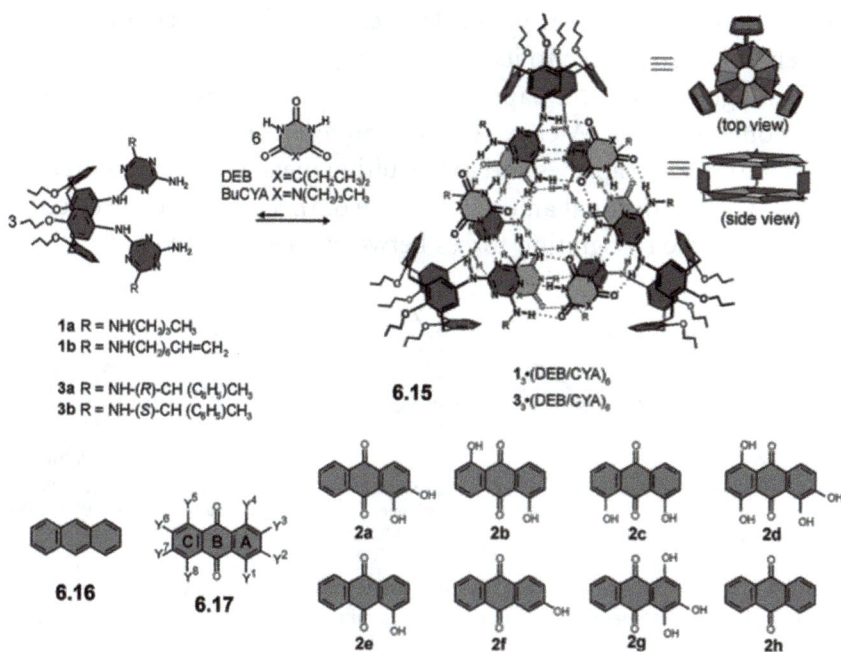

Figure 6.3 Assembly of chiral double rosettes **6.15**. The cyanuric acid–melamine hydrogen-bonding motif creates a box for binding three guests such as anthracene **6.16**. [Reprinted with permission from *J Am Chem Soc* **127**, 12697–12708. Copyright 2005, American Chemical Society.]

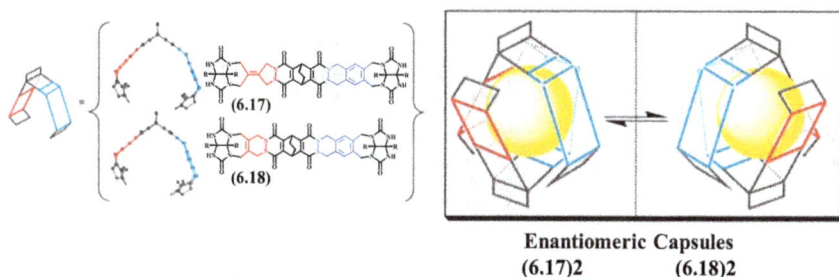

Enantiomeric Capsules
(6.17)2   (6.18)2

Figure 6.4 *Left*: Structures of desymmetrized modules. *Right*: Cartoons of the chiral softball capsules.

only C_2 axes, and chiral molecule guests can be expected to prefer one enantiomeric space to its mirror image.

Though racemic, the capsules could still be used to select between enantiomeric guests. A chiral guest is preferentially bound in one of the enantiomeric capsules and the resulting diastereomeric complexes are not present in equal amounts (Figure 6.5). The degree of selection depends on the recognition forces between guest and host.

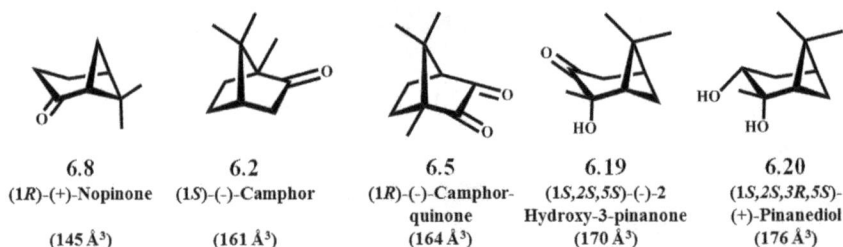

6.8	6.2	6.5	6.19	6.20
(1R)-(+)-Nopinone	(1S)-(-)-Camphor	(1R)-(-)-Camphor-quinone	(1S,2S,5S)-(-)-2 Hydroxy-3-pinanone	(1S,2S,3R,5S)-(+)-Pinanediol
(145 Å³)	(161 Å³)	(164 Å³)	(170 Å³)	(176 Å³)

Figure 6.5 Racemic softballs $(6.17)_2$ and $(6.18)_2$ are formed from modules bearing different spacers. Terpene guests showed some enantioselective binding (Table 6.1).

Table 6.1 Thermodynamic Encapsulation Data. K^1, Apparent Association Constants; $-\Delta G^0$, Free Energies of Formation; DE, Diastereomeric Excess; $\Delta(\Delta G^0)$, Calculated Differences in Stability between Diastereomeric Complexes of $(6.17)_2$ and $(6.18)_2$ with Chiral Guests in p-xylene-d_{10} at 295 K.

Guest	K'_A (M⁻¹)	$-\Delta G^0_A$ (kcal mol⁻¹)	K'_B (M⁻¹)	$-\Delta G^0_B$ (kcal mol⁻¹)	DE (%)	$\Delta(\Delta G^0)$ (kcal mol⁻¹)
		Complex $(6.17)_2$				
6.8	630	3.8	620	3.8	0	0
6.2	1100	4.1	870	4.0	12	0.1
6.5	960	4.0	850	4.0	6	0.1
6.19	1200	4.1	800	3.9	19	0.2
6.20	810	3.9	420	3.5	32	0.4
		Complex $(6.18)_2$				
6.8	290	3.3	270	3.3	0	0
6.2	420	3.5	300	3.3	17	0.2
6.5	310	3.4	250	3.2	12	0.2
6.19	300	3.4	170	3.0	29	0.4
6.20	390	3.5	190	3.1	35	0.4

These forces are expected to be amplified in the confined space, but are thwarted by the "breathing" motions of the capsule. Breathing occurs along the hydrogen-bonding seams that distort the shape space at relatively low energetic cost. The binding data reveal that the diastereoselectivity increases with the size of the guest and that the smaller capsule is somewhat more selective for a given guest. This is in accord with the anticipated effects of a snug fit between components, but the selectivities are moderate.

Nonracemates

A more robust version, **6.21**, was prepared to reduce the rates of racemization and slow the release of guests. This was accomplished through the addition of more hydrogen-bonding sites on the aromatic spacer unit (Figure 6.6). The capsule could then be manipulated[8] to leave an enantio-enriched imprint — a ghost — in solution. The key was to apply the observation that guest exchange is inevitably faster than capsule racemization. The exchange of guests can occur through opening of "flaps" of the softball, while racemization requires complete dissociation of halves (Chapter 1). The use of a chiral guest leaves its imprint on the capsule: one asymmetric cavity is preferred by the chiral guest. The rapid removal of that guest leaves a chiral but nonracemic set of (solvent-filled) host capsules that are stable for up to 20 h. This solution "remembers" the departed or "ghost" guest template (rather than its mirror image). In the simplest demonstration, Jose Rivera, Tomas Martin and Stephen Craig used (+) pinane diol **6.21** in *p*-xylene d_{10} with the capsule (Figure 6.7). This gave an equal

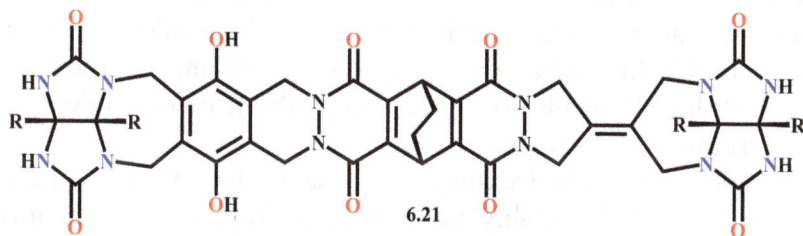

Figure 6.6   Additional (phenolic) hydrogen-bonding sites create a more robust capsule dimer (Figure 6.7).

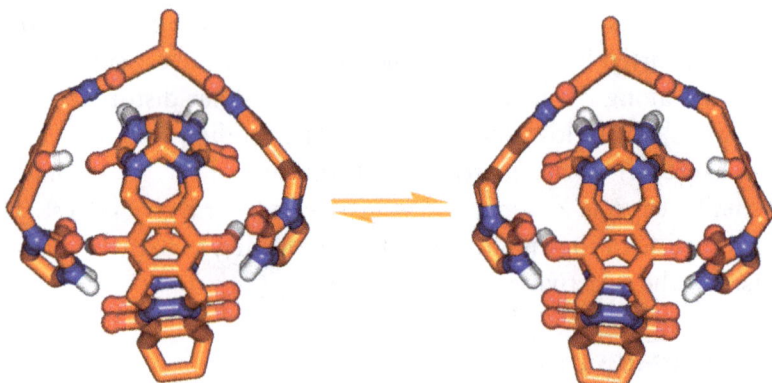

Figure 6.7 Racemization of $(6.21)_2$ involves complete dissociation and recombination of its halves. Guest exchange, however, is faster than racemization.

amount of its two diastereomeric complexes on mixing, but with time (a few days) an equilibrium was reached where a 50% excess of one complex was present. Excess (−) pinane diol **6.22** was then added and the *less stable* complex was temporarily formed in excess, which slowly re-equilibrated to the 50% diastereomeric excess state.

This proved a point in a convoluted way, but even more curious behavior could be elicited from the ghost. The final solution above could be evaporated and the excess pinanes were washed away with hexane, leaving a 2-to-1 mixture of (−) pinane in the enantiomeric capsules. The (−) pinane was exchanged with the achiral guest adamantanol **6.23**, and freed (−) pinane was again washed away. Now addition of (+) pinane displaced the adamantanol and gave the 2:1 mixture *of the less stable diastereomer* (Figure 6.8). This slowly equilibrated to the 2:1 mixture of the more stable diastereomer with time. In other words, the imprint of the chiral ghost was maintained by the achiral adamantanol. While imprinting on covalent materials has a long history,[9] these cases indicate that even dynamic systems can show the effects of memory.

Useful enantioselective complexation and chiral catalysis requires even higher kinetic stability and may profit from a combination of noncovalent and covalent interactions. So Rivera prepared a unimolecular version of the chiral softball (Figure 6.9).[10] This compound, **6.24**, resembles a clamshell but its lengthy synthesis (>25 steps) and

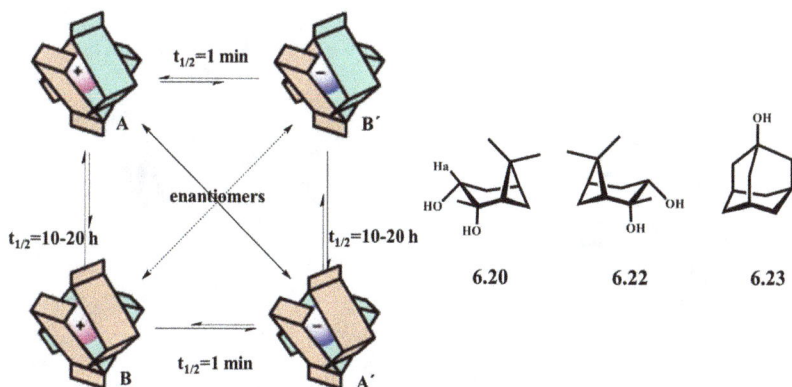

Figure 6.8 *Left*: Cartoons of racemization and guest exchange in the chiral (**6.21**)$_2$. *Right*: The enatiomeric guests and "ghost" maintaining adamantanol. [Reprinted with permission with *Angew Chem Int Ed Eng* 39(12), 2130–2132. Copyright 2000, Wiley-VCH, Weinheim.]

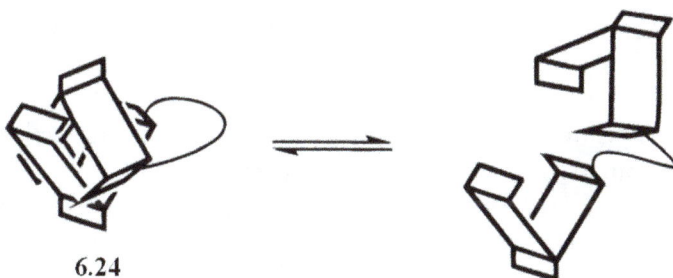

Figure 6.9 A cartoon depiction of a unimolecular softball — the clamshell.

modest diastereoselection (<20%) sharpened the contrast with, say, existing catalysts for asymmetric synthesis, and further development could not be justified.

Enantiopure Spaces

Optically Active Modules

The least ambiguous approach to chiral capsules involves the synthesis of modules that are actually chiral and single enantiomers.

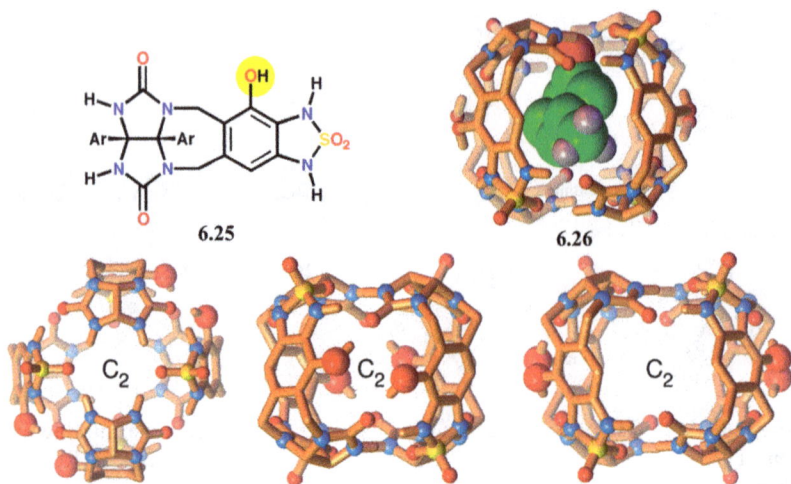

Figure 6.10   An optically active module **6.25** gives a tetrameric capsule with three C_2 axes. The assembly **6.26** with 3-methylcyclohexanone showed a 60% enantiomeric excess. [Reprinted with permission from *J Am Chem Soc* **121**, 10281–10285. Copyright 1999, American Chemical Society.]

The appropriate module **6.25** (Figure 6.10) was prepared by further desymmetrizing the tennis ball by Fraser Hof, Colin Nuckolls and Tomas Martin. As described earlier (Chapter 1), the curvature of the module distinguishes the molecule's convex front and concave back while the superior hydrogen bond donors (sulfamide) and acceptors (glycoluril) distinguish the right end from the left. This module assembles as an achiral tetramer with suitable guests such as adamantanedione. Introduction of a substituent (the hydroxy group) on the aromatic spacer now distinguishes the top of the module from its bottom. Having distinguished its three (informal) Cartesian coordinates, we see that the module is chiral. It was synthesized and the enantiomers were separated on a chiral Pirkle column. The assembly led to a chiral and enantiomerically pure space. The figure shows the shape of the capsule's space with views along its three C_2 axes and the complex with 3-methyl-cyclohexanone **6.26**. This guest showed the largest diastereomeric excess (60%).[11] The interaction of the ketone with the seam of hydrogen bonds places the methyl group of the favored guest enantiomer into the largest opening of the capsule host.

Optically Active, Sealed and Deep Cavitands

A series of unimolecular capsules **6.27** with a chiral "end" was prepared in enantiopure form by Shoichi Saito and Colin Nuckolls.[12] These molecules resembled deep cavitands but their upper rims were sealed with a seam of hydrogen bonds between hydroxyl groups on adjacent walls (Figure 6.11). The capsules were easily synthesized from the well-known resorcinarene platform **6.28** as the floor. The reaction with dihalo heterocyclic imide walls **6.29** gave the deepened cavitands. The requisite imides were prepared from condensation of the corresponding diacids with any number of chiral amino alcohols, including those derived from natural amino acids. The synthesis places the asymmetric centers on the "lid" of the container, where they are not ideally placed for interaction with the guest — rather like the case with the calixarenes. However, the disposition of the seam of hydrogen bonds **6.30** likely imparts a twist to the four walls, as shown in Figure 6.12, and transmits the asymmetry to the capsules' spaces.

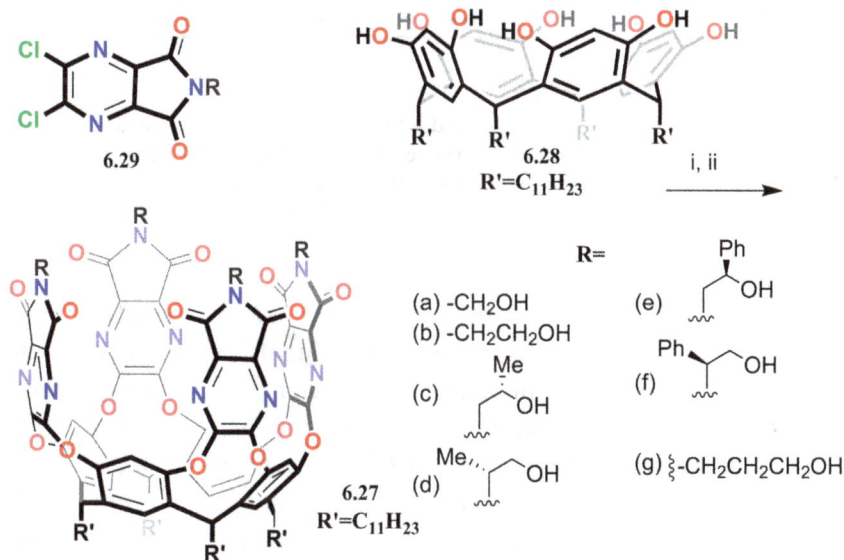

Figure 6.11   Synthesis of a cavitand with a chiral "lid."

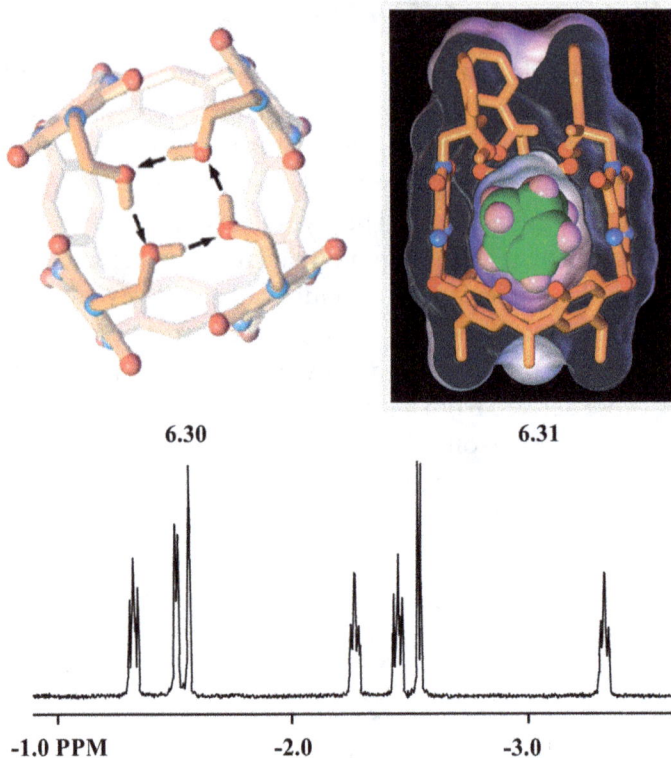

6.30   **6.31**

Figure 6.12   *Top*: The hydrogen-bonded lid **6.30** ensures slow exchange of guests, such as the complex **6.31** with norbornene. *Bottom*: The NMR spectrum of **6.31** shows the effects of the chiral microenvironment. [Reprinted with permission from *J Am Chem Soc* **122**, 9628–9630. Copyright 2000, American Chemical Society.]

To be sure, the magnetic environment is satisfyingly chiral, as shown by the spread of the NMR signals of norbornene as guest **6.31**. The steric influence of the chiral space was, again, modest with ee's comparable to those obtained earlier <35%. The application to separations proved practical, however, through the attachment of the capsule to a solid support by functionalizing the peripheral "feet."[13] The enantiopure cavitand derived form 2-amino-1-phenyl-ethanol was covalently bound to commercial polystyrene beads through a remote function. A clean separation of norbornene from adamantane in *p*-xylene solution could be accomplished using a common pipette loaded with the beads.[14] Likewise, the facile removal of trace methyl cyclopentane from 99.9% hexane was achieved in this way, and partial resolution of 1,2-trans-cyclohexane-diol enantiomers could be achieved in one pass through the beads.

Chiral Coencapsulation

The synthesis of chiral and enantiopure spaces, sedulously pursued, is an undertaking requiring considerable commitment of resources and a long attention span. Naturally, we were looking for short cuts, and one finally occurred to us: one way of creating a chiral space is to start with an achiral space, and place a chiral object into it. Take, for example, a cube-shaped **6.32** or spherical **6.33** room and introduce a single glove (Figure 6.13). The space around the glove now is chiral, but it is unlikely to make any difference in, say, enantioselection. Consider, now, shrinking the room to a much smaller volume — say, a volume just slightly larger than the glove. If the probe is a hand, then one would expect a good deal of interaction between hand and glove, possibly leading to selection between a right hand and a left hand groping for the glove. Ideally, the hand and glove would be arranged so that they can interact appropriately, and this requires a room with a shape space that can orient the two chiral objects; in short, a space of some different shape than a simple sphere or a cube. The cylindrical capsule was ideal for this type of molecular alignment since sizable molecules did not tumble freely in the space and rarely did molecules exchange positions while inside the capsule. Alessandro Scarso, Alex Shivanyuk and Osamu Hayashida reduced these new notions of chiral spaces to experimental practice.

For the molecular events a means by which two different compounds (rather than two copies of the same molecule) can be simultaneously encapsulated was required. The long cylindrical space inside the capsule was ideal for this purpose. As we have seen, molecules can be oriented along its axis and the capsule can select two different molecules that together fill its space properly. Consider styrene oxide or mandelic acid: with either of these guest one molecule does not fill

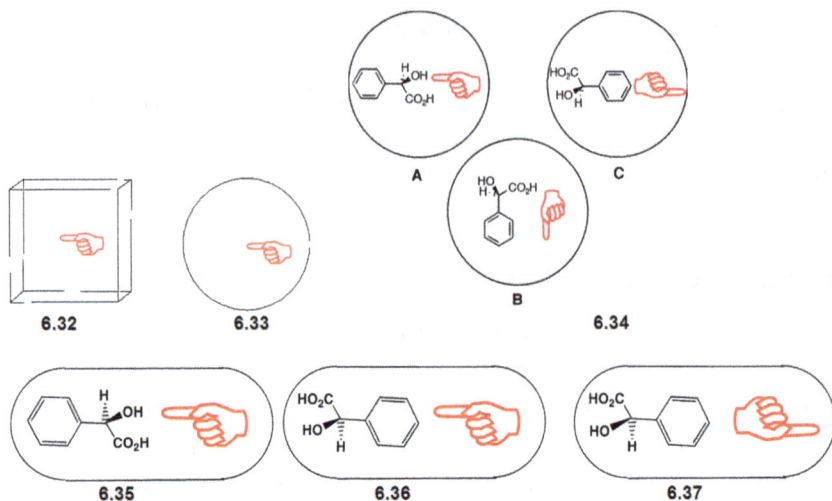

Figure 6.13   *Top*: A chiral object in an achiral box **6.32** or sphere **6.33**. Free tumbling as in **6.34** reduces the effect of the chiral space on the coencapsulated molecule. *Bottom*: Mandelic acid in the favored orientation **6.35** places a chiral center near the coecapsulated molecule.

enough of the space, while two molecules are too large to fit inside. Accordingly, they require coencapsulation with smaller molecules. The polar seam of hydrogen bonds favors arrangement **6.35** over **6.36** or **6.37**.

The first indication that magnetic effects of chirality were involved came during the coencapsulation of isopropyl chloride **6.38** with styrene oxide **6.39** (Figure 6.14).[15] The epoxide function was bound near the polar center of the capsule, presenting the chiral center to the co-guest, isopropyl chloride.

When these two molecules are free in solution, there are many possible contacts, interactions and orientations which are rapidly averaged. Consequently, the two methyls of the isopropyl group are equivalent (or at least nearly so) and they appear as a singlet in the NMR spectrum. This is because there are no particular attractions between **6.38** and **6.39** in dilute solution — styrene oxide is not an effective NMR shift reagent. However, in the confines of the cylindrical capsule, we find that the two methyl groups of **6.38** appear as *diastereotopic*, i.e. the asymmetric center creates a magnetic *and*

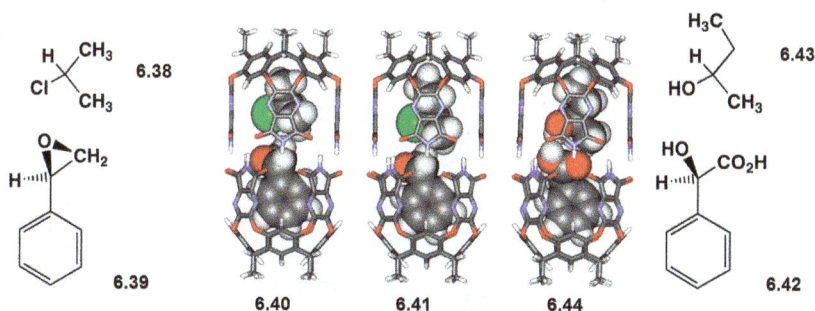

Figure 6.14 A chiral molecule in an achiral capsule creates a chiral space. An ee of about 70% was observed.

spatial anisotropy that influences a molecule held nearby, as in **6.40**. "Nearby" is the operative word here, because **6.38** is doomed to be constantly confronted with the asymmetric center of **6.39** in the small space of the capsule.

This propinquity is also the case when styrene oxide is in a related coencapsulated complex with 2-chlorobutane shown as **6.41**. When the halide co-guest is racemic, two different complexes are obtained with a modest diastereoselectivity. The modesty reflects the rather indifferent attractions that the two co-guests hold for each other. The highest selectivity that we saw was with a molecular pair that did have potential attractions. This was with mandelic acid and 2-butanol **6.43**. With this pair, both polar groups were expected to bind near the center of the capsule and they could also directly interact through hydrogen bonding as modeled in **6.44**. This apparently they did, and an enantiomeric excess (ee) of about 70% was observed. Now this level of enantioselectivity is certain to be regarded as failing by the standards of, say, catalytic asymmetric synthesis. There, contact with metal ions is involved and these are strong intermolecular forces. But we make no apologies: in a system held together with fragile hydrogen bonds, almost any enantiomeric excess (ee) is welcome, as it can foster development of candidates for organo-catalytic type applications.

Another expression of the effects of chiral centers at close range was uncovered in this capsule by Michael Schramm, Felix Zelder and Per Restorp.[16] In this case, a coencapsulated molecule **6.45** could

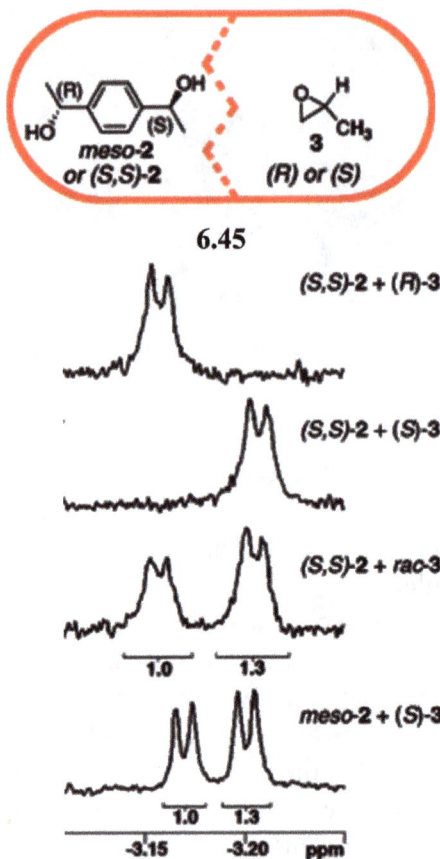

Figure 6.15 *Bottom*: Partial NMR spectra showing the differences in signals induced by remote stereocenters. *Top*: The diastereomeric complexes examined. [Reprinted with permission from *J Am Chem Soc* **125**, 1497–1499. Copyright 2008, American Chemical Society.]

"see" beyond a nearby chiral center to sense asymmetry at some distance, thanks to the intimate placement and orientation of the molecules in the cylindrical space (Figure 6.15).

What is the source of the differences sensed by the propylene oxide in **6.45**? This epoxide is confronted for a prolonged time with the steric and magnetic effects of the co-guest, fixed in a small space. We cannot rule out a slight (steric) distortion of the capsule walls that is transmitted from the remote center, but can say that this left no trace in the CD spectrum of the assembly. Instead, we believe that the magnetic effects of the remote center, fixed in time and space, are felt by the epoxide. Whatever the cause, the subtle effect resembles a report of xenon in a covalent cryptophane biosensor that differentiated an asymmetric center located seven covalent bonds away.[17]

The co-guests need not be so different to exert their stereochemical effects on each other. For example, in early experiments with 1,2-cyclohexane diol and the cylindrical capsule **6.46**, Thomas Heinz and Dmitry Rudkevich showed that the capsule prefers to take up a molecule and its mirror image rather than two molecules of identical chirality (Figure 6.16).[18] In free solution the different solute–solute interactions are often distinguishable by NMR,[19] but in the capsule these differences are amplified by the long contact times and infrequent exchange of partners. The capsule's preference is modest and cannot be generalized for other pairs of enantiomers. For example, Liam Palmer studied a series of carboxylic acids[20] that showed idiosyncratic behavior for halo carboxylic acids. Complex

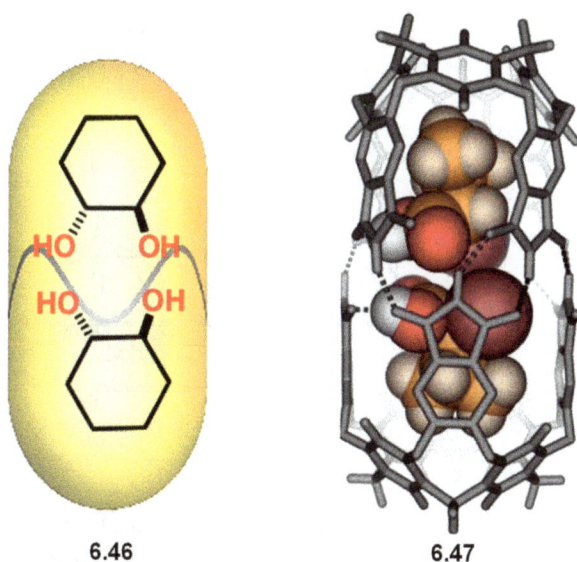

6.46   **6.47**

Figure 6.16   The capsule prefers mirror images of cyclohexane diols as in **6.46**, but identical molecules of α-Br-butyric acid as in **6.47**.

6.47 shows (*R*)2-Br-3-methylbutyric acid which indicated a small (1.5:1) preference for two identical guests. This preference was confirmed by calculations performed by our collaborators Yi-Lei Zhao and Ken Houk at UCLA.

Magnetic versus Spatial Effects of Chirality

The separation of steric and magnetic aspects of chirality is a rare event, and we conclude this section with another example involving our capsules. Toru Amaya synthesized a series of capsules **6.48** with remote asymmetric centers *outside* the cavities,[21] as shown in Figure 6.17. None of the capsules showed CD traces, despite the

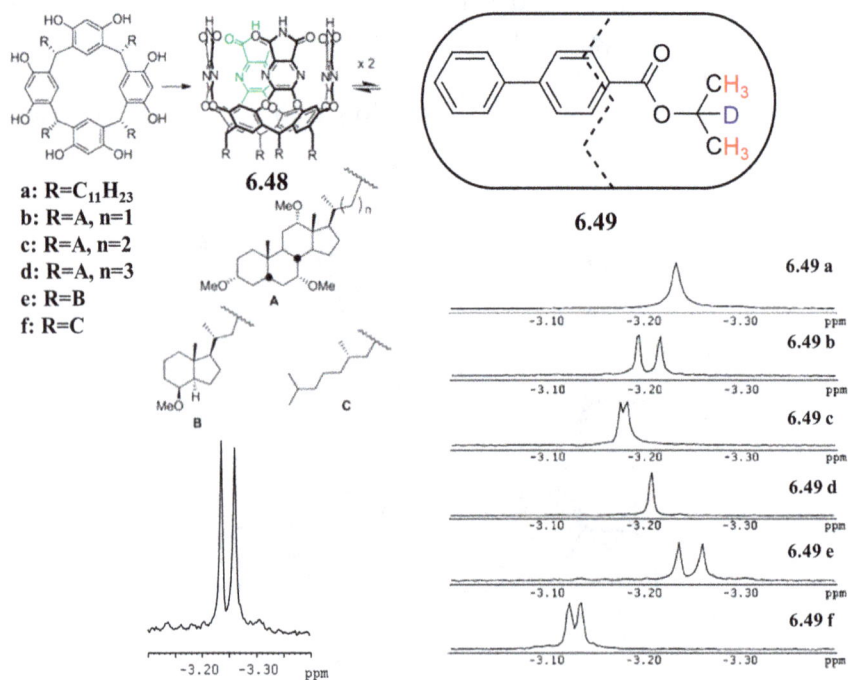

Figure 6.17 Cylindrical capsules with chiral centers outside. Although the space inside is achiral, the isopropyl groups of the ester inside **6.49** sense the asymmetric magnetic environment outside and show diastereotopic methyl signals in the NMR spectra. [Reprinted with permission from *J Am Chem Soc* **126**, 6216–6217. Copyright 2004, American Chemical Society.]

many asymmetric centers (88 in the cholic acid derivatives). In other words, the space inside the capsules was not chiral. Yet several of the capsules **6.49** showed the effects of the exterior asymmetric centers on the NMR spectrum of the guest. This guest is rigidly fixed in the space and maintains its isopropyl groups at the end of the cavity. When asymmetric centers are 3–4 carbons away (but not farther), it shows diastereotopic methyl groups in the NMR spectrum. Clearly, the magnetic aspects of chirality can penetrate the mechanical barriers represented by the capsule walls. Since there is no possible contact between the asymmetric centers outside and the i-Pr groups inside, this system has separated the steric and magnetic aspects of chirality.

The sensing of asymmetry through a mechanical barrier raises the possibility of remote control of stereoisomerism in capsules. For example, radical pair reactions such as those that result in chemically induced dynamic nuclear polarization (CIDNP) might be influenced by a remote asymmetric center.[22]

References

1. Costante-Crassous J, Marrone TJ, Briggs JM, *et al.* (1997) Absolute configuration of bromochlorofluoromethane from molecular dynamics simulation of its enantioselective complexation by cryptophane-C. *J Am Chem Soc* **119**, 3818–3823.
2. Tokunaga Y, Rebek Jr J. (1997) Chiral capsules: softballs with asymmetric surfaces bind camphor derivatives. *J Am Chem Soc* **120**, 66–69.
3. Castellano RK, Kim BH, Rebek Jr J. (1997) Chiral capsules: asymmetric binding in calixarene-based dimers. *J Am Chem Soc* **119**, 12671–12672.
4. Timmerman P, Verboom W, van Veggel FCJM, *et al.* (1994) A novel type of stereoisomerism in calix[4]arene-based carceplexes. *Angew Chem Int Ed Eng* **33**, 2345–2348.
5. Corbellini F, Di Costanzo L, Crego-Calama M, *et al.* (2003) Guest encapsulation in a water-soluble molecular capsule based on ionic interactions. *J Am Chem Soc* **125**, 9946–9947.
6. Kerckhoffs JMCA, ten Cate MGJ, Mateos-Timoneda MA, *et al.* (2005) Selective self-organization of guest molecules in self-assembled molecular boxes. *J Am Chem Soc* **127**, 12697–12708.
7. Rivera JM, Martin T, Rebek Jr J. (1998) Chiral spaces: dissymmetric capsules through self-assembly, *Science* **279**, 1021–1023.
8. Rivera JM, Craig SL, Martín T, Rebek Jr J. (2000) Chiral guests and their ghosts reversibly-assembled hosts. *Angew Chem Int Ed Engl* **39** 2130–2132.

9. See for example: Polborn K, Severin K. (1999) Molecular imprinting with an organometallic transition state analogue. *Chem Commun* **24**, 2481–2482 and references therein.
10. Rivera JM, Rebek Jr J. (2000) Chiral space in a unimolecular capsule. *J Am Chem Soc* **122**, 7811–7812.
11. Nuckolls C, Hof F, Martín T, Rebek Jr J. (1999) Chiral microenvironments in self-assembled capsules. *J Am Chem Soc* **121**, 10281–10285.
12. Saito S, Nuckolls C, Rebek Jr J. (2000) New molecular vessels: synthesis and asymmetric recognition. *J Am Chem Soc* **122**, 9628–9630.
13. Saito S, Rudkevich DM, Rebek Jr J. (1999) Lower rim functionalized resorcinarenes: useful modules for supramolecular chemistry. *Org Lett* **1**, 1241–1244.
14. Saito S, Rebek Jr J. (2001) Synthesis and application of a deep, asymmetric cavitand on a solid support. *Bioorg Med Chem Lett* **11**, 1497–1499.
15. Scarso A, Shivanyuk A, Hayashida O, Rebek Jr J. (2003) Asymmetric environments in encapsulation complexes. *J Am Chem Soc* **125**, 6239–6243.
16. Schramm MP, Restorp P, Zelder F, Rebek Jr J. (2008) Influence of remote asymmetric centers in reversible encapsulation complexes. *J Am Chem Soc* **130**, 2450–2451.
17. Spence MM, Ruiz EJ, Rubin SM, *et al.* (2004) Development of a functionalized xenon biosensor. *J Am Chem Soc* **126**, 15287–15294.
18. Heinz T, Rudkevich DM, Rebek Jr J. (1999) Molecular recognition within a self-assembled cylindrical host. *Angew Chem Int Ed Engl* **38**, 1136–1139.
19. See for example: Jursic BS, Goldberg SI. (1992) Enantiomer discrimination arising from solute–solute interactions in partially resolved chloroform solutions of chiral carboxamides. *J Org Chem* **57**, 7172–7174.
20. Palmer LC, Zhao YL, Houk KN, Rebek Jr J. (2005) Diastereoselection of chiral acids in a cylindrical capsule. *Chem Commun* **29**, 3667–3669.
21. Amaya T, Rebek Jr J. (2004) Steric and magnetic asymmetry distinguished by encapsulation. *J Am Chem Soc* **126**, 6216–6217.
22. Muus LT, Atkins PW, McLauchlan KA, Pedersen JB, eds., (1977) *Chemically Induced Magnetic Polarisation*. D. Reidel, Dordrecht.

CHAPTER 7

Expanded and Contracted Capsules

As we have described in earlier chapters, some shapes — helical and folded alkanes — are unknown in solution; and even enzyme interiors show only bent alkyl structures.[1] Tight-fitting alkane chains appear in other synthetic assemblies[2] and covalent complexes.[3] Their contortions and tight packing are now becoming advantageous in catalysis.[4,5]

Expanded Capsules

The advantages that a nonspherical space offers for guest alignment led us to search for expanded versions of the cylindrical capsule **7.1** (Figure 7.1). Attempts to extend the walls or the resocinarene base by incorporating naphthalene units in place of benzenes met with solvophobic collapse,[6] and made it unlikely that covalent means of extension would succeed.[7] Instead, a simple module inserted between the capsule's halves promised minimal synthetic effort. Consider diimides, for example durene diimide **7.2**. It provides the appropriate hydrogen-bonding complementarity but its limited solubility did not allow insertion as in **7.3**, and other diimides fared no better.

The solution came with Dariush Ajami's application of the glycoluril modules **7.4** (Figure 7.2). Their bent shapes (though not at the ideal right angles of the capsule's walls) and their wealth of hydrogen-bonding donors and acceptors were expected to complement the imides as shown. The patterns of the donors and acceptors on glycolurils complement the adjacent walls of the cavitand; the ureido carbonyls are superior hydrogen bond acceptors and the acidic N–H's of the imides are the best donors. The slight divergence of

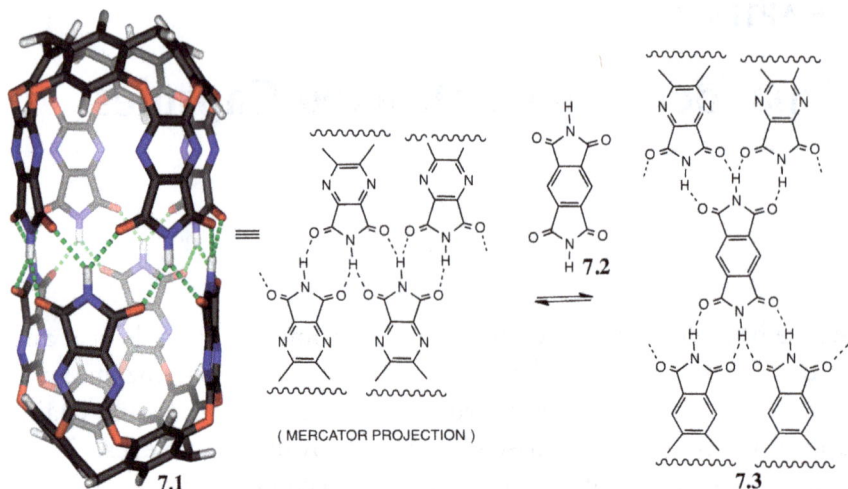

Figure 7.1 The original cylindrical capsule **7.1** and its attempted expansion by insertion of durene diimide units **7.2**.

the capsule's walls was not expected to be a problem, and it was not. Insertion occurred, but not in the predicted geometry or stoichiometry.[8] A "belt" of *four* glycolurils was inserted into the middle of the capsule in a twisted arrangement that created a chiral assembly **7.6**. This structure is known hereafter as the *expanded capsule*, but to enhance solubility the aryl functions of the glycolurils were modified to bear various peripheral groups (*p*-octyloxy, *p*-dodecyl or *p*-dibutylamino) that are not always specified in the structures below.

As indicated in structure **7.6**, a total of 24 hydrogen bonds hold the new assembly together. Each glycoluril makes four hydrogen bonds with the cavitands and four hydrogen bonds to adjacent spacers. The strongest hydrogen bonds are between the imide N–H of the cavitand and the ureido oxygen of the glycoluril. This leaves one carbonyl oxygen of each imide wall without a hydrogen bond donor. The steric clashes of these unpaired carbonyl groups with the adjacent imide panels are relieved by "twisting" the array of the imide walls as shown in Figure 7.3. The twist changes the symmetry of the cavitands from C_{4v} to C_4, and produces a chiral structure. The chiral (but, of course, racemic assembly) emerges from achiral subunits.[9]

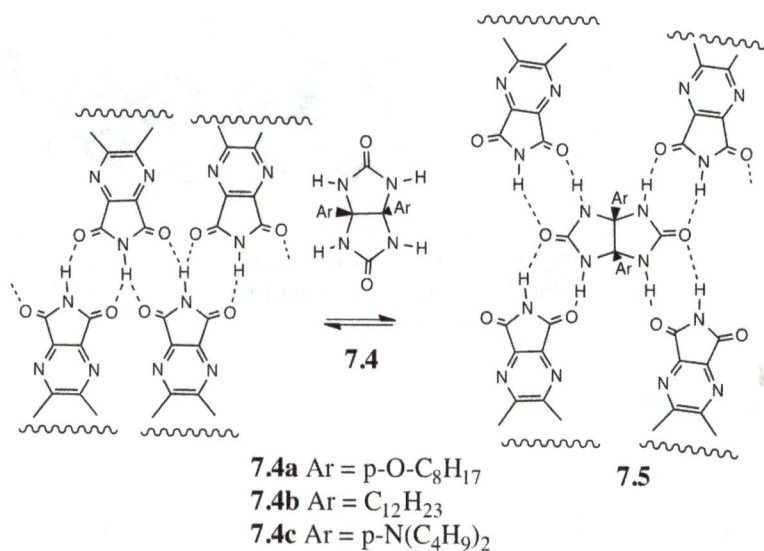

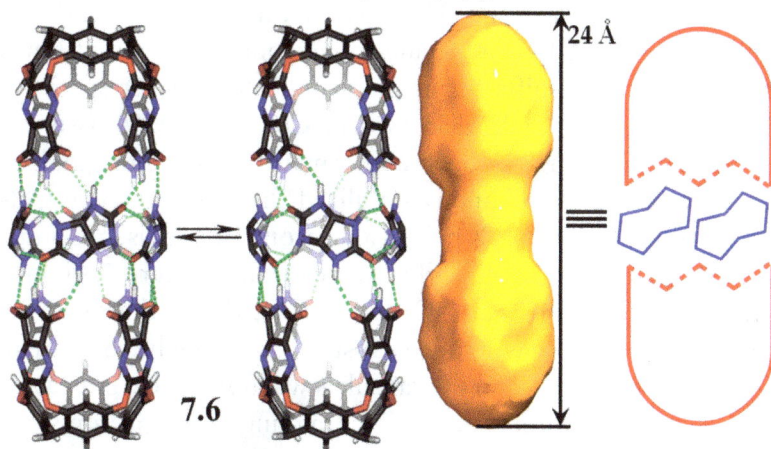

Figure 7.2 *Top*: Expansion of the capsule with glycoluril spacers **7.4** was expected to create capsule **7.5**. *Bottom*: Instead, the racemic assembly **7.6** was formed (the peripheral alkyl groups of the capsule and the aryl groups of the glycoluril have been removed for clarity). The cartoon representation used elsewhere in this chapter is also shown.

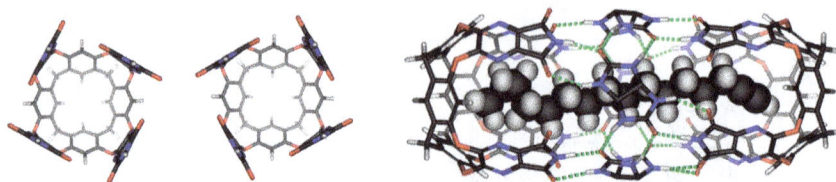

Figure 7.3 *Left*: Twisting of the walls of the capsule creates an asymmetric magnetic environment. *Right*: Energy-minimized structure of the complex of **7.6** with 1-hexadecyne. [Reprinted with permission from *J Org Chem* **74**, 6584–6591. Copyright 2009, American Chemical Society.]

Encapsulation

Alkanes

The consequences of the capsule expansion became immediate on encapsulation of tetradecane. The NMR spectrum of C14 in the new assembly shows that four molecules of glycoluril are present but the changes in chemical shifts of the guest indicate a longer space than in the original capsule. As discussed in Chapter 3, C14 assumes a coiled shape with many *gauche* conformations in the original capsule, where the coiled alkane is stabilized through attractive CH–π interactions with the cavity's polarizable aromatic panels.[10] Comparison of the signals of C14 in the two capsules (Figure 7.4) shows that the hydrogens on C_2/C_{13}, C_3/C_{12} and C_4/C_{11} of this guest have shifted downfield in the expanded capsule. Accordingly, these hydrogens have moved away from the capsule's walls and ends, indicating a relaxed, extended conformation. The doubling of the proton signals on C_2/C_{13} and C_4/C_{11} in the longer capsule reflects an asymmetric magnetic environment; these hydrogens are diastereotopic.

A Spring-loaded Molecular Device

The two states of encapsulated C14 — compressed and extended — were reminiscent of a spring-loaded device, and Dariush Ajami devised a means of controlling them. He prepared a glycoluril **7.4c** bearing peripheral basic sites. It featured good solubility in mesitylene and was readily inserted into the original cylindrical capsule to

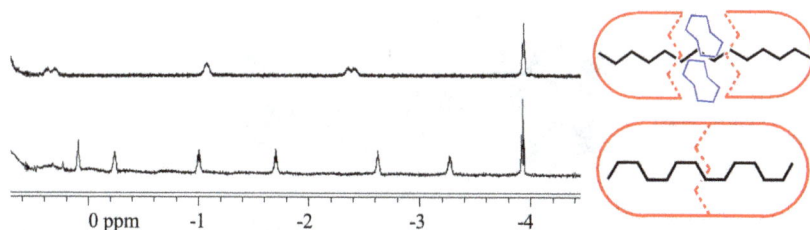

Figure 7.4 *Top*: Upfield portion of the ^{1}H NMR spectrum (600 MHz, mesitylene-d_{12}) of C14 (tetradecene) in expanded capsule **7.6**. The alkane is extended and in an asymmetric environment. *Bottom*: The same guest in the original capsule exists in a compressed helical conformation. [From *Proc Natl Acad Sci USA* **104**, 16000–16003. Copyright 2007, National Academy of Science of the United States of America.]

give the expanded capsule containing extended C14.[11] Addition of HCl gas into the NMR sample protonated the anilines of the glycoluril, which precipitated as its HCl salt (Figure 7.5). This left the original capsule with the compressed alkane inside. Then trimethylamine was added to the NMR tube. This deprotonated the glycolurils, which dissolved and reassembled the expanded capsule with C14 inside. Some six cycles of addition of acid and then base were possible in a single NMR tube before the buildup of trimethylamine hydrochloride deteriorated the ^{1}H NMR spectra. Accordingly, the extension and compression of C14 is controlled by access to the glycoluril spacers, which is controlled by the presence of acids and bases.

Compressed Longer Alkanes

Insertion of the glycolurils increases the length of the assembly by 7 Å and its inner volume by 200 Å^3, and the accommodation of *n*-alkanes from C14 to C19 is possible.[12] The longer guests also undergo compression to fit within the extended capsule. As the alkanes coil, the *gauche* conformations increase and force the hydrogens of the methylene groups nearer to the twisted walls. Darish Ajami used nucleus independent chemical shift (NICS)[13] computations to map the magnetic shielding/deshielding provided by the extended capsule. A depiction of the map obtained through calculations at the B3LYP/6–31G*

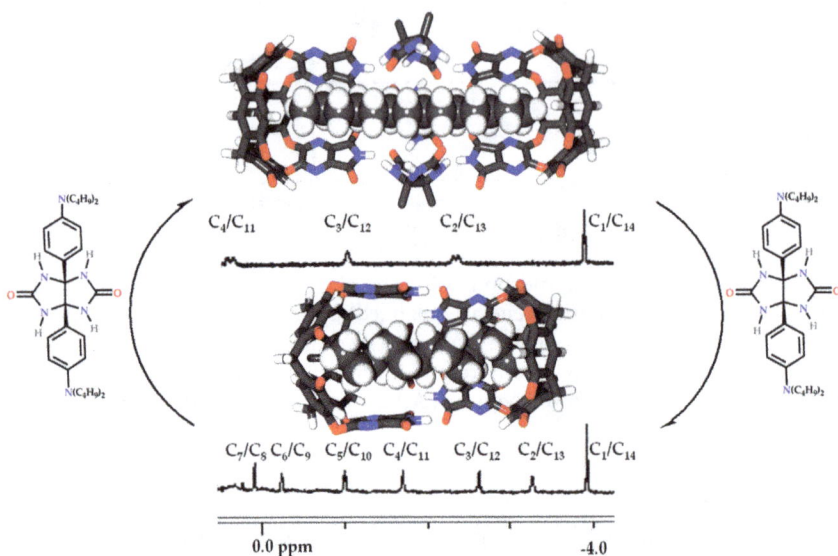

Figure 7.5 Encapsulated C14 is reversibly compressed and extended by the action of acids and bases. Acidic conditions protonate and precipitate the glycoluril, leaving coiled C14 inside the shorter capsule. Under basic conditions the glycolurils dissolve and insert to give the longer capsule with extended C14 inside.

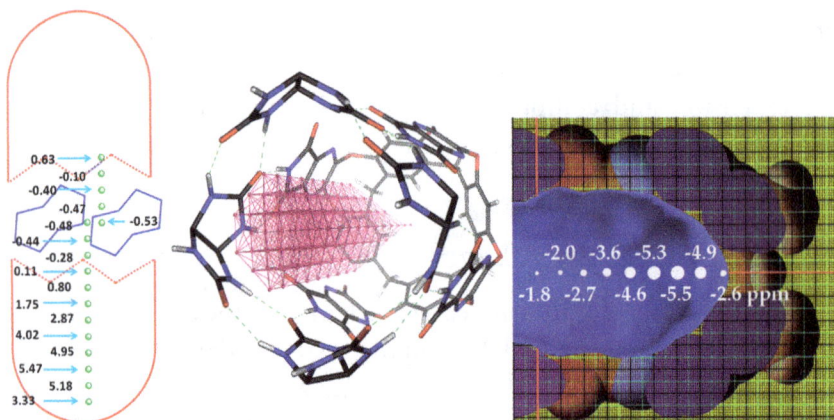

Figure 7.6 *Left* and *Right*: Map of the inner space with calculated NICS values along the central axis at 1 Å intervals. *Center*: Positions of atoms placed throughout the entire space during computations. [Reprinted with permission from *J Org Chem* **74**, 6584–6591. Copyright 2009, American Chemical Society.]

level of density functional theory[14] is shown in Figure 7.6. The values reported are those along the central axis at spacing distances of 1 Å. The values increase as nuclei move nearer the walls. The map can be used to predict the location of a guest nucleus inside the cavity, since the calculated values are in good agreement with the observed NMR shifts.[15]

Nuclei positioned near the middle of the extended capsule experience some deshielding as the four glycolurils present the edges of their eight aromatic units on the outer surfaces to guests inside.[16] For example, the vinyl hydrogens of *trans*-7-tetradecene are located near the center of the cavity and these signals are shifted *downfield* by a $\Delta\delta$ of $+0.72$ ppm. The methyl groups are located at the ends of the capsule and are shifted *upfield* by a $\Delta\delta$ of -4.7 ppm.

The coiled alkanes apply stress to the inside of the extended capsule, and the capsule squeezes guests. We were able to detect and evaluate these forces by studying the racemization of the capsules as a function of guest length. The racemization results in the coalescence of the diastereotopic NMR signals of the geminal hydrogens as the samples are heated. The rates were obtained by EXSY spectra taken at different temperatures, and the free energies of activation $\Delta G^{\ddagger}$ are readily calculated. The proposed racemization mechanism involves the concerted rotation of all glycolurils by $\sim30°$ in one direction (Figure 7.7). This process forms new hydrogen bonds at the expense of old ones as a slightly longer (achiral) intermediate is reached.[17] The more compressed guests apply more pressure to the interior of the capsule and force the assembly toward the longer, achiral intermediate. Accordingly, longer guests increase the racemization rates. Dariush Ajami found activation energies $\Delta G^{\ddagger} = 17.2$, 16.7 and 15.7 kcal/mol for C17, C18 and C19, respectively. The compressed conformations of these alkanes are modeled in Figure 7.7. In contrast, the alkane C16 can fit inside in a fully extended conformation, and no evidence of racemization was observed, even on heating.

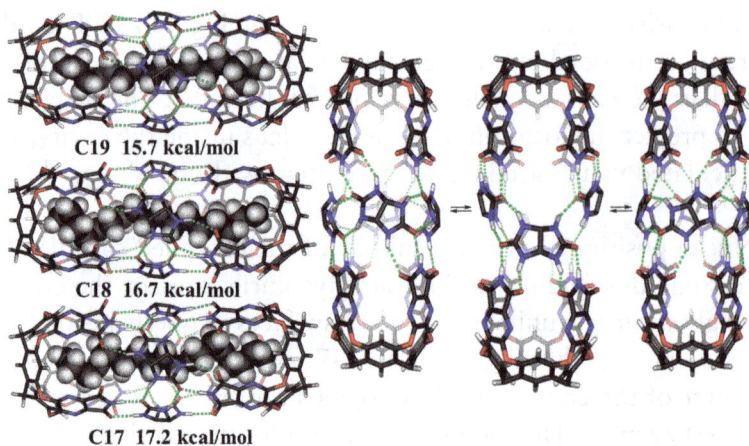

Figure 7.7 *Left*: Modeled alkanes and the activation energies for capsule racemization; the longer alkanes exert more pressure on the capsule and speed the racemization. *Right*: The proposed racemization mechanism — as the spacer units are rotated in a concerted manner, a longer and achiral intermediate is reached.

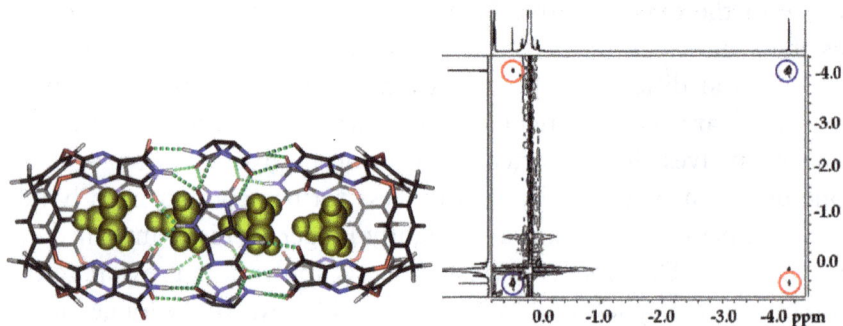

Figure 7.8 *Left*: Model of four encapsulated cyclopropanes. *Right*: Cross-peaks (circled in red) of a 2D ROESY spectrum reveal that guest positions can exchange on the NMR timescale. [Reprinted with permission from *Angew Chem Int Ed Eng* 47(32), 6059–6061. Copyright 2008, Wiley-VCH, Weinheim.]

Compressed Gases

One of the smallest guests encapsulated was cyclopropane and four molecules were taken up (Figure 7.8). Although separate, sharp signals for the two positions of the guest appear in the NMR spectrum, a 2D

ROESY spectrum showed that the encapsulated cyclopropanes can exchange their positions inside the capsule on the NMR timescale.[18] An activation barrier of 18.5 kcal/mol for the exchange of positions was calculated from integration of the cross-peaks. The cyclopropanes fill about 36% of the space and application of ideal gas laws to the assembly gives a pressure of several hundred atmospheres although ambient pressure was used to load the gases. Obviously, ideal gas laws do not apply: the cyclopropanes are not point masses, nor are their collisions with the walls elastic. Instead, attractive CH–π interactions between the guests and the aromatic panels lower the energy of the system and permit these high "pressures." It is even worth asking if "phase" is an appropriate term when only a few molecules are involved.

Compressed Alkynes

The shape of the space inside, particularly the tapered ends and the constricted center of the capsule, can select between various functional groups of the guests. This is illustrated by a model of encapsulated C16 acetylene, as shown in Figure 7.9.[19] Specifically, a narrow terminal acetylene can access the space of the resorcinarene, but the larger methyl "knob" of an alkane cannot penetrate this space as deeply. The constricted center of the assembly requires a thin, extended alkane conformation, and the remote end of the alkane must buckle to be accommodated. The methylene NMR signals of encapsulated C16 acetylene near the methyl end move physically closer to the walls of the capsule and are shifted upfield (Figure 7.9).

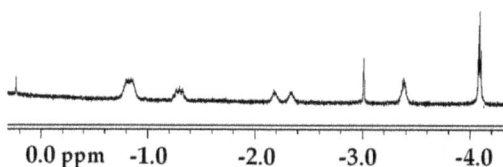

Figure 7.9   Upfield regions of the ^{1}H NMR spectrum (600 MHz, mesitylene d_{12}) of 1-hexadecyne in the expanded capsule. The acetylenic C–H (-3.0 ppm) penetrates the resorcinarene, while the methyl end must coil to be accommodated.

Compressed Hydrogen Bonds

A study of hydrogen-bonded carboxylic acid dimers in the expanded capsules gave evidence of compression in the isolated environment. This research was undertaken in collaboration with the Limbach group in Berlin, who have developed sensitive probes for H-bond geometries.[20–23] Correlations between ^{1}H NMR chemical shifts, $O \cdots O$ distances and hydrogen positions have been established using solid state NMR, X-ray diffraction and neutron scattering.[24–28] Dariush Ajami and Henry Dube encapsulated longer carboxylic acid dimers in the expanded assembly and saw slow H-bonded proton exchange. The chemical shifts of the $O \cdots H–O$ signals moved downfield as the effective length of the dimers increased: 14.59 ppm for *p*-toluic acid, 14.77 ppm for *p*-ethyl-benzoic acid, 15.98 ppm for *p-tert*-butyl-benzoic acid and 16.72 ppm for *p*-methyl-cinnamic acid (Figure 7.10).[29] Spectra were also obtained at intermediate deuterium fractions that showed separate signals for the HH dimer and the HD

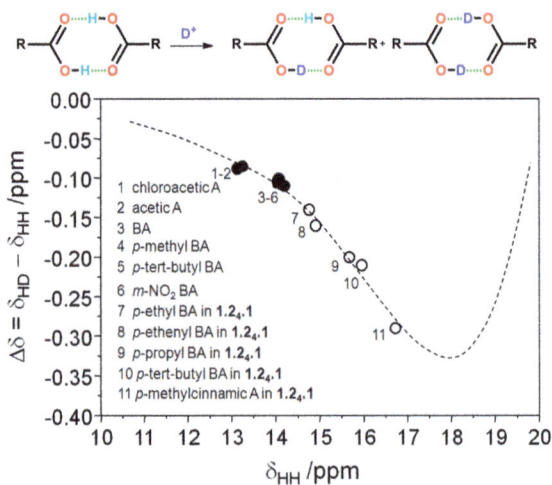

Figure 7.10 Plot of H/D isotope effects, $\Delta\delta = \delta_{HD} - \delta_{HH}$, for the benzoic acid (BA) derivatives in the expanded capsule (d_{12} mesitylene — open circles) and in solution (CDF_3/CDF_2Cl — filled circles) as a function of the bridging proton chemical shift δ_{HH}. [Reprinted with permission from *Angew Chem Int Ed Eng* 50, 528–531. Copyright 2011, Wiley-VCH, Weinheim.]

dimer. The difference, defined as $\Delta\delta = \delta_{HD} - \delta_{HH}$, serves as a measure of pressure. The encapsulated carboxylic acid dimers behave as if they were under considerable pressure from the inner walls of the capsule. The estimated pressures were 4–10 kbars, in agreement with the amplification of intermolecular forces during the temporary isolation of species in capsules.

The behavior of the series of carboxylic acids discussed above underscores the effects of the capsule's tapered ends. By any external measure, the *p*-ethyl, *p*-isopropyl and *p-t*-butyl acids have the same length, but they show different positions — and different "effective sizes" — in these capsules. A *p*-ethyl group cannot penetrate very far into the resorcinarene, and the rest of the molecule moves off the capsule's central axis. The CH_2 is closer to the twisted walls and experiences the chiral environment. Only one methyl of an isopropyl group can be in the resorcinarene at a given time, and isopropyls show diastereotopic methyls in their NMR spectra. A *tert*-butyl group cannot access the tapered space at all and is pushed away from the ends. The best fit for the tapered spaces at the ends is methyl groups that are located on the central axis (e.g. tolyl).

Social Isomerism of Guests

The extended capsule has enough space to allow two *p*-disubstituted benzenes to fit inside. When these guests are unsymmetrical, a number of different arrangements — social isomers — are possible.[30] We introduced social isomers in Chapter 3 as a consequence of the shape space of the capsule that maintains the arrangements of guests. Specifically, the guests are too long to tumble freely inside the space and too thick to slip past each other on the NMR timescale. In the extended capsule, *p*-cymene, *p*-ethyl-toluene and *p*-methyl-styrene were convenient sources of social isomerism[15] (Figure 7.11).

Simple monosubstituted phenyl groups, ethyl, propyl or butyl benzenes always present the thinner alkyl groups toward the center of the expanded assembly. This space is constricted and cannot easily accommodate two benzene rings. Little evidence of social isomers is shown by these simple compounds. Likewise, for the *p*-cymene case

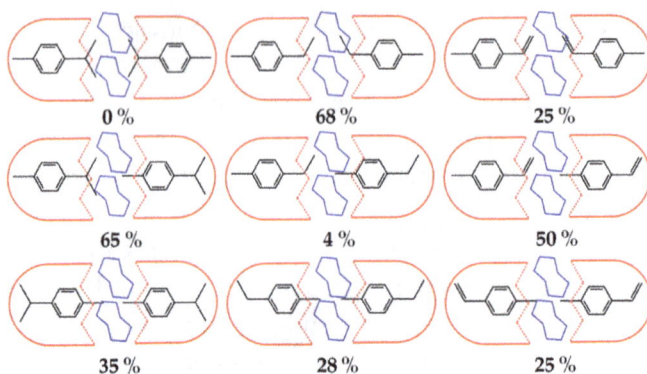

Figure 7.11   Social isomers of (*left*) *p*-cymene, (*center*) *p*-ethyl-toluene and (*right*) *p*-methyl-styrene in the expanded capsule. [Reprinted with permission from *Angew Chem Int Ed Eng* **50**(39), 9150–9153. Copyright 2011, Wiley-VCH, Weinheim.]

only two of the three possible isomers are observed as two isopropyl groups are an uncomfortable fit in the center. However, with *p*-ethyl-toluene the most favored isomer has both ethyl groups in the center and, unexpectedly, the statistically favored unsymmetrical isomer represents only a few percent of the mixture. In contrast, *p*-methyl-styrene shows exactly what would be expected from a statistical distribution of social isomers. We developed further applications of the expanded capsule, as its ability to position co-guests in predictable orientations is unique among hydrogen-bonded capsules.[31–40] In fact, few capsules of any sort — covalently bonded,[41,42] assembled with metal–ligand interactions,[43,44] or even hydrophobic effects[45] — have the capacity to position co-guests. Ajami and Dube evaluated hydrogen-bonding interactions between boronic acids, carboxylic acids and primary amides in the expanded capsule.[46] The phenyl boronic acids are recent components of covalently self-assembled systems,[47] and we found that the *p*-methyl, methoxy, ethyl and isopropyl derivatives all fit as symmetrical dimers in the expanded capsule. The structure of the boronic acid dimer has been a subject of debate, but a recent theoretical study found the doubly hydrogen-bonded exo/endo conformer (Figure 7.12) to be lowest in energy.[48] The structure has planar (C_{2h}) symmetry and features both involved and spectator hydrogens.

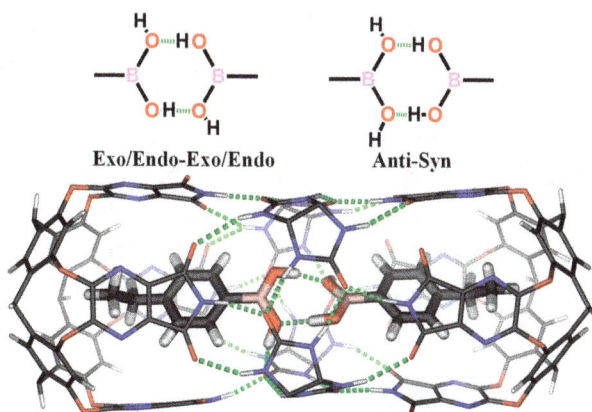

Figure 7.12 *Top*: Computed structures of the H–B(OH)$_2$ hydrogen-bonded dimers; *exo–endo* and *anti–syn* isomers are accessible. *Bottom*: The energy-minimized structure (HF/6–31 Ag*) of the *exo–endo* isomer of the *p*-ethyl-phenyl boronic acid dimer in the expanded capsule. [Reprinted with permission from *Angew Chem Int Ed Eng* 50(39), 9150–9153. Copyright 2011, Wiley-VCH, Weinheim.]

*Know abbreviation for calculations.

In more polar media, the alternate syn/anti dimer was energetically accessible.

The spectator acidic hydrogens accelerate the racemization of the capsule, and other guests can catalyze the rotation of the glycolurils. Any hydrogen bond donor and acceptor groups at the middle of the assembly can speed up the interconversion of the capsule enantiomers through acid/base catalysis. This effect is evident in broadened NH signals of the glycoluril spacer, and indicates reasonably rapid racemization of the capsule on the NMR timescale. Ajami and Dube examined combinations of encapsulated carboxylic acids, boronic acids and carboxamides to determine the strongest interactions in the context of the capsule. Direct competition experiments between guests of the same size (the *p*-ethyl derivatives) were used to eliminate the effects of "fit."

All three pairwise combinations and their respective (less probable) homodimers were observed. The boronic acid homodimer was always favored, whether it competed with the carboxylic acid, the amide or both functions. The relative stabilities inside the expanded capsule were: boronic acid homodimer > acid/amide hetereodimer > acid/boronic acid heterodimer > acid/acid homodimer > boronic acid/amide hetereodimer > amide homodimer. The simultaneous characterization of all these species is impossible in solution, because the rapid exchange of partners averages the NMR signals. However, reversible encapsulation allows the dissection of these equilibria and showcases the power of this technique in qualitative physical organic chemistry.

Hyper-extended Capsules

Single Molecule Guests

The expanded capsule could accommodate normal C15, C16, C17, C18 and C19, but the appearance of a new hyper-expanded capsular assembly in the presence of C19 was a surprise. Ajami determined that this capsule incorporated *two* belts of glycolurils[49] and showed that it accommodated lengthier alkanes. It came as no surprise then that longer hydrocarbons such as C24–C29 induced the formation of an even longer complex involving *three* belts of glycoluril spacers (Figure 7.13). The encapsulation results are reported in Table 7.1. A total of 15 molecules make up the hyper-extended assembly, but evidence of even longer capsules exists. The solubility of relevant glycolurils and the formation of gels have hampered their characterization. The simple recipe of the cavitand, glycoluril and long alkanes delivers a banquet of structures with a range of guest conformations. Lijuan Liu and Ajami termed these mutual but unanticipated arrangements of hosts and guests "soft templates."[50] The unique flavor can be recognized as follows: neither component — say, the coiled alkane nor the hyper-expanded capsule — exists on its own, but they do coexist.

One of the dividends paid by reversible encapsulation equilibria was to arrive at a rule of the proper filling of space — the 55% solution.[51] The filling of space is an obvious feature of other recognition phenomena, even in those synthetic containers that do not completely surround their targets,[52] and those naturally occurring compounds like enzymes and receptors. Simply filling space characterized some of

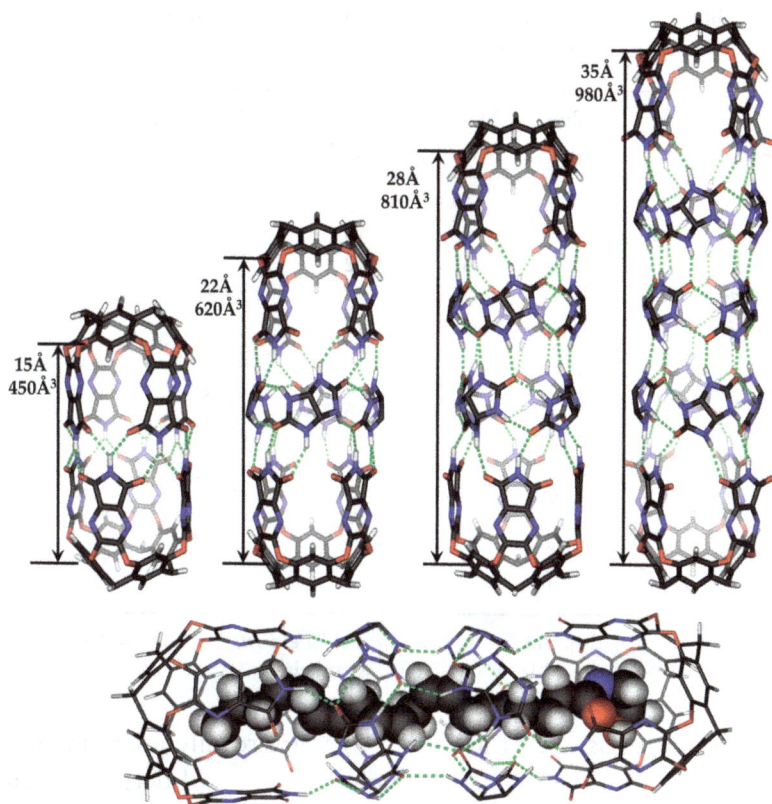

Figure 7.13 *Top*: Energy-minimized structures and approximate dimensions of capsules extended by glycolurils. The length refers to the accessibility of methyl groups in the inner space. *Bottom*: The energy-minimized (AM1) structure of encapsulated anandamide inside the doubly expanded capsule. Peripheral groups have been removed for viewing clarity.

the earliest, finite self-assemblies based on melamine/cyanuric acid recognition in solution.[53] These complexes had no other function. However unconventional, our approach has always looked for departures from mainstream physical organic chemistry,[54] even when it leads to uncharted waters and causes anxiety among the lifeguards.[55] The hyper-extended capsules provided the required tools for encapsulating more complex and even bioactive compounds, and a test of how well the 55% solution held up. The adaptability of a number of natural products having long and narrow shapes has been

Table 7.1 Relevant Data for Alkane Dimensions and their Packing Coefficients (PC's) are given.

Guest	Volume (Å^3)	PC (%) in 1.1	PC (%) 1.2$_4$.1	PC (%) in 1.2$_8$.1	PC (%) in 1.2$_{12}$.1
n-C$_{13}$H$_{28}$	230	54	37		
n-C14H$_{30}$	247	58	40		
n-C$_{15}$H$_{32}$	264	62	42		
n-C16H$_{34}$	281		45		
n-C$_{17}$H$_{36}$	297		48		
n-C$_{18}$H$_{38}$	314		51	39	
n-C19H$_{40}$	331		53	41	
n-C$_{20}$H$_{42}$	348		56	43	
n-C$_{21}$H$_{44}$	364			45	
n-C$_{22}$H$_{46}$	381			47	
n-C$_{23}$H$_{48}$	399			49	41
n-C$_{24}$H$_{50}$	417			51	43
n-C$_{25}$H$_{52}$	434			54	44
n-C$_{26}$H$_{54}$	450				46

[Reprinted with permission from *Angew Chem Int Ed Eng* **46**, 9283–9286. Copyright 2007, Wiley-VCH, Weinheim.]

probed and it has been found that they are readily taken in by the cavitand/glycoluril system. For example, fatty acid derivatives such as anandamide (Figure 7.13), which is the endogenous ligand for the cannabinoid receptor of the brain,[56] are encapsulated.[57] Its subunit, arachidonic acid, is the substrate for the biosynthesis of the prostaglandins, and several other signaling molecules, including capsaicin, are bound. Oleamide, one of the fatty acid amides involved in sleep induction,[58] and its ethyl ester derivative were also encapsulated. For future applications in medicine, the controlled release of these natural products from the capsules would have to be arranged and the stoichiometric nature that makes these assemblies too expensive must be overcome. Another hurdle in the medical application process is establishing these capsules as transporters across membranes, but these aspects are outside the scope of the present treatise. A brief discussion of release is given in a subsequent section below.

Ion Pairs

The interaction of acids and bases in solution gives rise to a number of short-lived and rapidly interconverting species. But their isolation in capsules makes interactions a prolonged and private matter, and the expanded systems seemed likely to capture ion pairs. Toshi Taira studied mixtures of trifluoroacetic acid (TFA) and picoline which gave symmetrical assemblies with a signal far downfield (18.7 ppm) for the acidic proton in the NMR spectrum.[59] This is a region where strong hydrogen bonds appear, indicating that the acidic proton is in contact with both the picoline nitrogen and the TFA oxygen. Moreover, since this proton is coupled to the α C–H of the picoline, it must be transferred to the nitrogen. The stoichiometry corresponds to the hyper-expanded capsule: two of each component (cavitand, picoline and TFA) and eight glycolurils are involved (Figure 7.14).

Unexpectedly, the methyl signals are shifted downfield; the CH_3 occurs at 3.8 ppm, whereas methyl signals in the ends of the assembly would appear in the upfield region (<0 ppm). This signifies that the orientation of the picolines is reversed from the usual: their methyls are near the assembly's center by the glycolurils. This conclusion is confirmed by the ^{19}F NMR spectrum, which shows a signal at -83.5 ppm. This signal for the CF_3 groups is shifted upfield by nearly 6 ppm from the free acid, and puts the TFA's in the deepest ends of the container assembly.

Similar spectra were obtained for 4-ethylpyridine with TFA and for γ-picoline with CF_3SO_3H. In the latter, the acidic proton in this case was not shifted as far downfield. With the less acidic difluoroacetic acid (DFA) the NMR signals reveal instead a H-bonded complex with γ-picoline inside. No coupling to the heterocycle's α C–H is seen and the downfield signal for the acidic proton disappears. Diastereotopic F signals are observed in the ^{19}F NMR spectra and expose the chiral nature of this assembly, which arises from the twisted arrangement of the spacers. The assembly racemizes on heating, with coalescence of the diastereotopic ^{19}F signals. The isolated encapsulation complexes show behavior quite different from those of molecules in solution or in the solid state.[60]

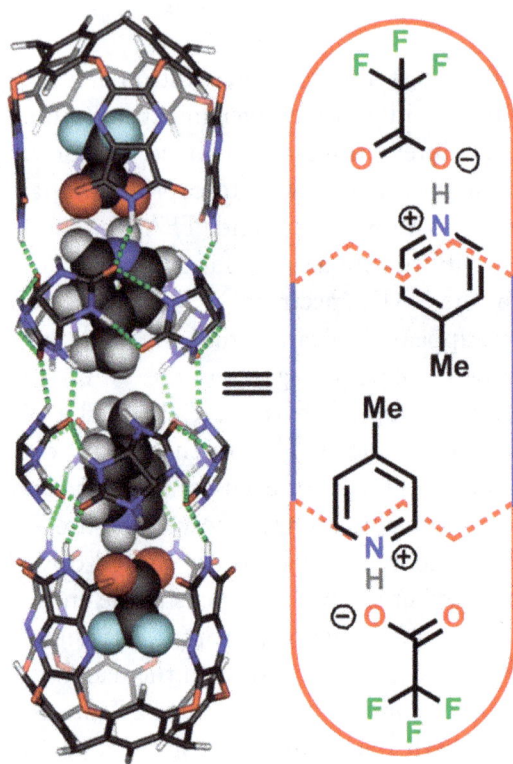

Figure 7.14   Coencapsulation of ion pairs.

Reactivity

Another anomalous behavior of an encapsulated agent was discovered by Taira and Ajami. The *n*-octylisocyanate is coencapsulated with benzene in the expanded capsule as established by the DOSY spectrum.[61] This solution maintained at 294 K gradually forms the symmetrical N,N-dioctylurea — but this product appears in the hyperexpanded capsule (Figure 7.15)! Control experiments established that water enters the assembly and hydration takes place inside the extended capsule, facilitated by the glycoluril spacers. Evidence for the resulting carbamic acid as the intermediate was obtained by NMR, and insertion of another belt of spacers is induced by the entry of

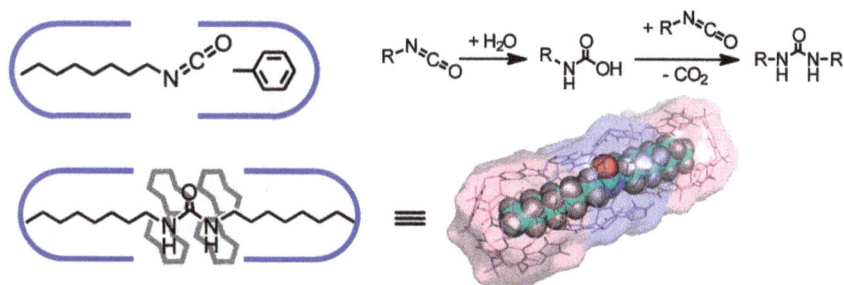

Figure 7.15 *Top*: Coencapsulation of an isocyanate and benzene in an extended capsule. *Bottom*: The hydration with adventitious water leads to a symmetrical urea in a hyper-extended capsule.

a second isocyanate. The sequence leading to the urea product is shown.

Photocontrol of Guest Exchange

The exchange of molecules in and out of capsules is required for applications in catalysis, and Henry Dube applied the well-established system that uses light: the *cis–trans* photoisomerization of azobenzenes.[62,63] The azobenzenes had been used in proteins,[64,65] but in supramolecular chemistry they were first applied in crown ethers[66–68] and then cyclodextrins.[69,70] The photoisomerization predictably changes the shape, and we used it in an indirect sense to control the access of other guests to encapsulation.

In Chapter 3 we described how the isomerization controlled access to the original cylindrical capsule. Irradiation gives the *cis* isomer, which forces out walls into a partial vase / partial kite conformation. The incoming guest flushes out the ill-fitting *cis* azobenzene. Brief heating at 160°C reconverts the azo compound to its *trans* conformation, which rapidly replaces the resident guest. The irradiation–heating cycle can be repeated many times without deterioration of the spectra. Parallel experiments were performed with the expanded capsule, using the appropriately longer azobenzene *trans*-4-methyl-4'-hexyl-azobenzene[71] as the guest (Figure 7.16).

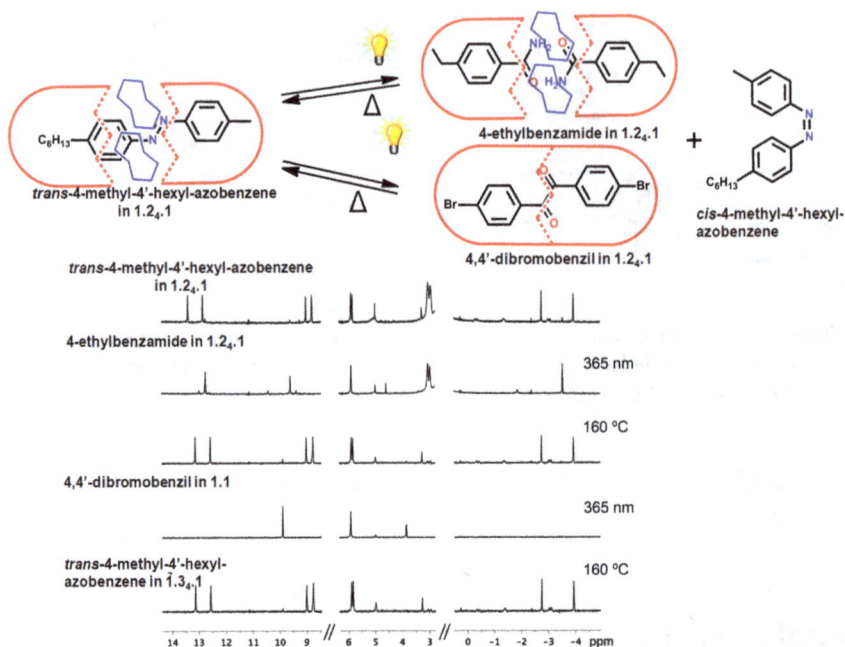

Figure 7.16   Light-induced guest exchange and assembly rearrangement. Replacement of *trans*-azobenzene by dimeric 4-ethylbenzamide maintains the expanded assembly. With 4,4′-dibromobenzil as the incoming guest, light induces the formation of the shorter capsule. [Reprinted with permission from *Angew Chem Int Ed Eng* **49**(18), 3192–3195. Copyright 2010, Wiley-VCH, Weinheim.]

Competition experiments with 4-ethylbenzamide show only the encapsulated azo compound. The relevant regions of the ^{1}H NMR spectrum are shown and separate signals for the two cavitand ends are seen for the unsymmetrical azo guest. After irradiation at 365 nm, the hydrogen-bonded homodimer of 4-ethylbenzamide replaced the azo compound as the guest. This new assembly's symmetry is reflected in its simplified NMR spectrum. Heating the sample to 160°C for 2 min restored the initial state, and the cycle could be repeated many times.[72]

Dube also used photoisomerization as a means of switching between the original cylindrical capsule and the expanded one. The solubility of *p*-dodecylphenyl glycoluril (**7.4b**, Figure 7.2) in deuterated mesitylene is low, but it dissolves when incorporated into the expanded assembly with *trans*-4-methyl-4'-hexyl-azobenzene.[73] The azo compound is the only guest even in the presence of added 4,4'-dibromobenzil. When the mixture is irradiated for 50 min at 365 nm, only the shorter capsule with 4,4'-dibromobenzil as guest is observed. The disproportionation is aided by the precipitation of the glycouril from solution. The original extended assembly can be restored on heating the suspension for 2 min at 160°C. Again, this cycle could be repeated many times to establish the photochemical control of expanded capsule versus shorter capsule assemblies.

A Different Spacer for Expansion

New Capsules

As mentioned earlier, the glycolurils were intended to fit into the corners of the cavitand and offer superior hydrogen bond acceptors to the imides' N–H donors. The twisted arrangement of walls in Figure 7.3 arises, in part, from a mismatch: the adjacent walls of the cavitand are at right angles (Figure 7.17) but the ureido functions of the glycoluril are folded at an angle of 113°.[74] Konrad Tiefenbacher surmised that a more appropriate complement to the cavitand would be a propanediurea (PDU). Its fold angle is ~99°,[75] and we expected

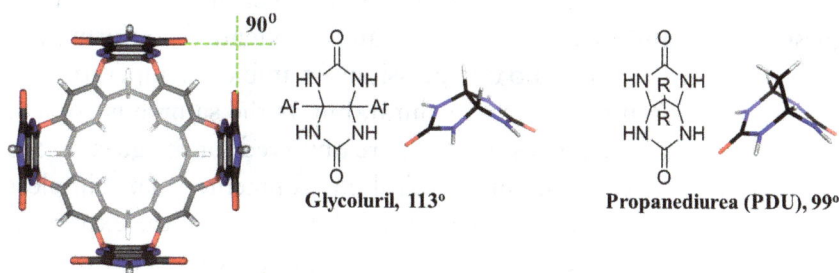

Glycoluril, 113°   **Propanediurea (PDU), 99°**

Figure 7.17 View of the cavitand and structures of glycoluril and propanediurea spacer modules. Peripheral alkyl and aryl groups are deleted. [Reprinted with permission from *Angew Chem Int Ed Eng* 50(50), 12003–12007. Copyright 2011, Wiley-VCH, Weinheim.]

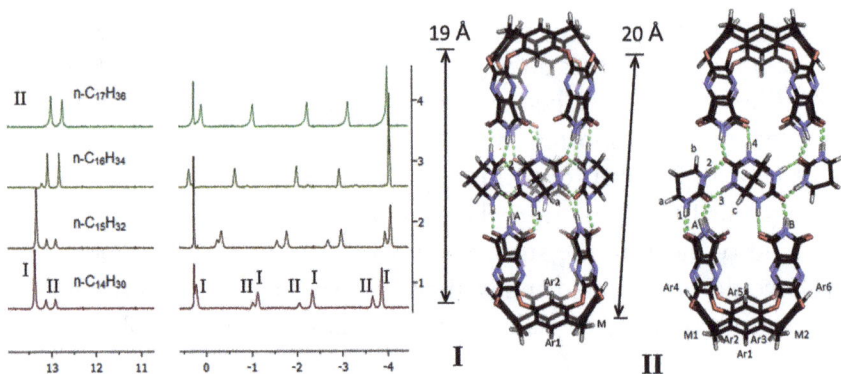

Figure 7.18 *Left*: ^{1}H NMR spectra (280 K) of guests from *n*-C14 to *n*–C17 in capsules expanded with propanediureas. The proposed arrangements of spacers in the two isomeric capsules. [Reprinted with permission from *Angew Chem Int Ed Eng* 50(50), 12003–12007. Copyright 2011, Wiley-VCH, Weinheim.]

its smooth insertion into the cylindrical capsule. But we were unprepared for the new "S" and "banana" shapes that emerged.

We relied on *n*-alkanes as guest probes since their characteristic NMR signals reveal their positions and conformations in the cavitand subunits. The shortest alkane, *n*-C14, gave *two* new complexes (labeled **I** and **II**; Figure 7.18, line (1), and no sign of the coiled conformation was found in the original capsule of the dimeric cavitand. Both new capsules incorporated four molecules of the PDU, and the symmetry of the guest signals indicated that the two ends of either capsule have the same magnetic environment. The chemical shifts of the guest signals, and their spacing, indicate an extended conformation of the guest with little or no compression. At ambient temperature or higher, both assemblies appear achiral, but as the sample is cooled, diastereotopic geminal guest protons are observed for the guest in the major assembly (**I**), indicating a chiral environment.[76] The simplest structure for the chiral assembly is the PDU version of the expanded capsule with a D_4-symmetric structure **I** (Figure 7.18).

The *n*-C14 guest inside the minor assembly **II** behaves as if it were experiencing a longer space, and the guest's methylene protons are not diastereotopic, even when the temperature is lowered: this extended capsule is an achiral structure. When the sample is heated, the two assemblies interconvert, as shown by coalescence of the signals at approximately 300 K. Encapsulation of the longer n-C15 showed similar NMR spectra. A ROESY experiment with this guest showed exchange between the guest signals in the two different assemblies. Molecular modeling helped us arrive at a C_{2h}-symmetric and S-shaped assembly for the structure of **II** (Figure 7.16). There are two kinds of spacers and the C_{2h} symmetry results in four different ureido N–H resonances and three bridgehead C–H signals of the PDU.

For the cavitand, two imide N–H signals, two different methine C–H's and six aromatic C–H signals appear. These are labeled in the figure. Many of the signals are also enantiotopic because a plane of symmetry exists in this assembly. The structure suggests an alternate intermediate for the racemization of the glycoluril expanded capsule. Instead of an intermediate with two planes of symmetry (shown earlier in Figure 7.7), the structure proposed for **II**, with PDU's replaced by glycolurils, would also explain the racemization mechanism. The racemization of the new capsule is faster and suggests a weaker hydrogen-bonded network with the PDU's.

The enantiotopic signals of the S-shaped capsule **II** were revealed by encapsulation of a chiral guest. We applied Waldvogel's approach; he had earlier[77] shown that 2-tetradecanol in the original cylindrical capsule induced local stereoselective helical folding. Doubling of the enantiotopic host signals of **II** occurred with the guest, 2-heptadecanol, and the imide N–H signals of the host became diastereotopic.

Hydrocarbons are soft templates that can compress and lengthen, depending on the available space. Likewise, the capsules can incorporate or shed spacers to accommodate the shape of the alkane: these systems display a mutually induced-fit behavior. An extended n-alkane chain increases the length by 1.25 Å for each methylene added but a coiled chain increases the length by 0.93 Å.[78] It was no surprise, then, that the longer n-C18 induced the emergence of yet a new encapsulation assembly **III** (Figure 7.19). Compared to its shape in **II**, the guest is in a more extended conformation, and separate imide NH signals are seen for the cavitands. Only **III** is observed with n-C19 and a chiral structure is required since a diastereotopic CH_2 group of

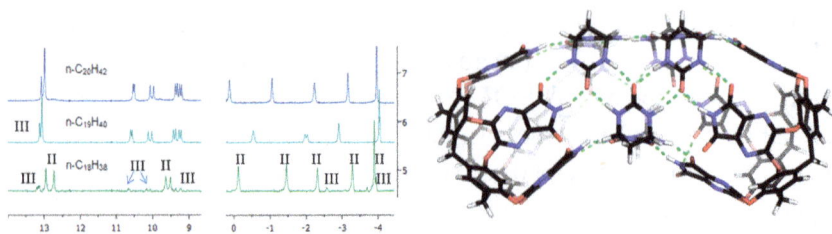

Figure 7.19 *Left*: NMR spectra of encapsulated C18–C20. *Right*: The modeled shape of the "banana" capsule assembly **III**. Peripheral alkyl groups have been deleted. [Reprinted with permission from *Angew Chem Int Ed Eng* 50(50), 12003–12007. Copyright 2011, Wiley-VCH, Weinheim.]

this guest can be observed at 240–300 K. Unprecedentedly, *six* spacer units of the PDU spacer are taken up; all other extensions of the parent capsule involve additions of multiples of four units. We propose the unusual banana-shaped structure (Figure 7.19) of C_{2v} symmetry, formed by the insertion of two PDU units into assembly **II**. The NMR spectrum shows four imide N–H's, four different methine C–H signals and eight aromatic C–H signals for the cavitands. For the PDU spacers, 12 different urea NH resonances and six bridgehead C–H signals were present.

Tiefenbacher supported the structure by NOE signals observed in the NMR spectra, by a simulated NMR spectrum at the DFT level of theory (B3LYP/6–31G*) and by extensive molecular modeling. Experimental (functional) evidence for the unusual shape of **III** came from its shape-selective encapsulation of rigid guests. Specifically, the rigid and linear *p*-pentaphenyl (Figure 7.20) was taken up by the doubly extended capsule described previously. It is inappropriately shaped for the curved **III** and was not encapsulated. On the other hand, the bent dialkynylketone was encapsulated in the congruent **III** but not in the inconveniently shaped linear capsule derived from glycolurils.

With *n*-C20 as guest, the same assembly **III** was present but the guest signals were shifted upfield, indicating a more compressed guest conformation. But the slightly longer *n*-C20 as guest induced the emergence of yet another new capsular species of the PDU. The new assembly coexists with **III**, but with the longer *n*-C22 only a single complex (**IV**; Figure 7.21, trace 9) emerged. Integration

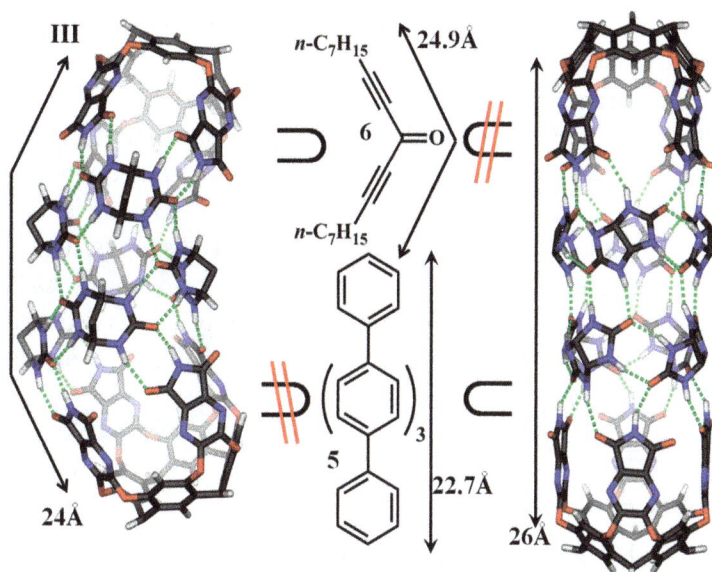

Figure 7.20   Selective encapsulation of complementary-shaped guests in assembly **III** and the previously reported, doubly extended cylindrical capsule.

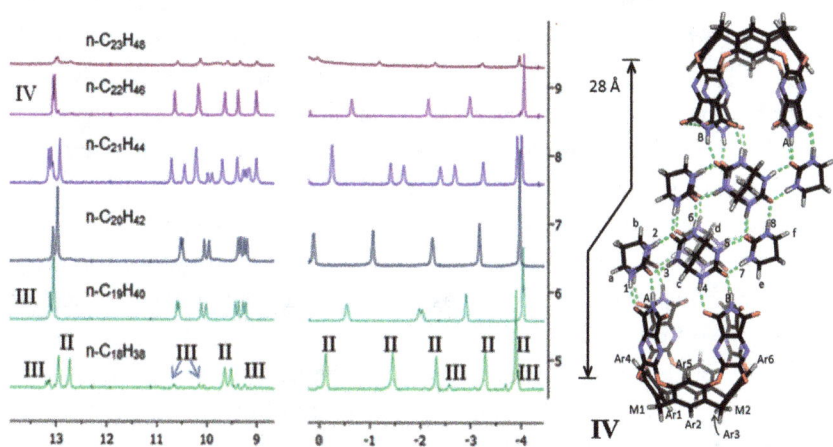

Figure 7.21   *Left*: Partial NMR spectra of increasingly long guests. *Right*: The proposed structure of a guest-induced, *S*-shaped capsule. [Reprinted with permission from *Angew Chem Int Ed Eng* 50(50), 12003–12007. Copyright 2011, Wiley-VCH, Weinheim.]

revealed that *eight* PDU units were present in the new assembly. The guests' methylene signals did not become diastereotopic even at low temperatures, indicating an achiral structure for this doubly extended capsule. We propose the D_2-symmetric structure for **IV**, and the proposal is supported by ^{1}H NMR data and results of NOE experiments. The cavitands feature two imide N–H signals, two different methine C–H signals and six aromatic C–H signals. The PDU's show eight different NH resonances and six bridgehead C–H signals. With the longer *n*-C23 the same assembly **IV** persists, but the signal-to-noise ratio is poor even when 60 equivalents of guest is present.

Aromatic Attractions

We found that a number of individual π–π interactions can impart selectivity to the self-assembly process of the new expanded capsules. Although these forces are weak, they can be responsible for an entirely different type of capsule,[79] and their sheer numbers in protein interiors contribute to protein folding and thermal stability.[80] For the case at hand, the coexistence of two complexes such as **I** and **II** with guests C14–C16 indicates that they are of comparable free energy. The assemblies represent two-state systems similar to Wilcox's torsion balance.[81] His ability to measure edge-to-face interactions suggested that even feeble aromatic π–π attractive interaction forces could tip the thermodynamic balance for the formation of one assembly over the other. The balance has found wide applications[82] for weak forces,[83] so we set our sights on measuring π–π interactions in the context of capsule stability.

Tiefenbacher noticed that the structure of the PDU units allows aromatic panels on its "bridge" to be oriented in two different ways. Direct attachment of phenyl groups to the framework (as in spacer **7**) results in aromatic units that diverge and are directed *away* from assembly **I** ((a) in Figure 7.22).[84] On the other hand, connection by way of an intervening CH$_2$ unit (as in the benzyl groups of spacer **8**) allows the aromatic panels to fold back on each other and even onto the outer surface of capsule **I**. The assembly's hydrogen-bonding array

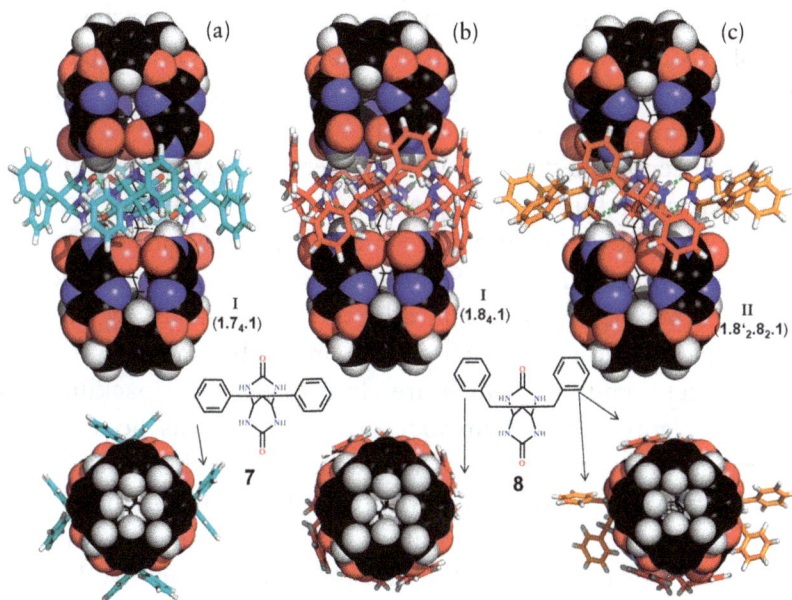

Figure 7.22 Models of assemblies **I** and **II** with different aromatic propanediurea spacers. *Left*: The phenyls of the spacer are directed away from the capsule's surface. *Center*: In assembly **I** the spacers are arranged in a "twisted" fashion (red). The phenyls can fold onto the capsule's surface. *Right*: In assembly **II** the "twisted" spacer (red) can fold but the "horizontal" spacer (orange) cannot. Assembly **II** has fewer stabilizing π–π interactions with the assembly's hydrogen-bonding array than does **I**. (Peripheral alkyl groups have been deleted for easier viewing.) [Reprinted with permission from *J Am Chem Soc* 134, 2914–1917. Copyright 2012. American Chemical Society.]

should be stabilized by such π–π interactions ((b) in Figure 7.22). Molecular modeling calculated the distance between the aromatic π surface and the hydrogen-bonding array to be ~3.5 Å, in agreement with literature precedents.[85–87]

In the capsule arrangement **I**, all eight benzene rings (on each of the four spacers) can fold inward and hide the seam of hydrogen bonds from the solvent ((b) in Figure 7.22). In making these contacts the benzenes *become* the solvent. But in assembly **II** half of these interactions are lost, since two of the spacers cannot stack their aromatics back onto the hydrogen bonding seam ((c) in Figure 7.22).

As a functional consequence, PDU's outfitted with the benzyl spacers induced only assembly **I** when exposed to the shorter guests (*n*-C14–C16). Even the longer C17 was bound exclusively in assembly **I**. Control experiments with benzyl spacers having bulky peripheral groups that precluded the approach showed that stacking was the driving force for the assembly.

But a deeper question is: What governs the structural changes in the two assemblies? The answer seems to be capsule capacity. Longer guests inevitably apply pressure on the two ends of any capsule. The compressed guests favor spacer orientations that increase the capsule's dimensions. Specifically, the accessible cavity length in **II** is about 1 Å longer than in **I**. An inevitable factor is the filling of space: complexes which are close to the ideal packing coefficient of slightly more than 50% are favored. This holds over a wide range of guest lengths. The induced fit of the host and the adaptability of the guest are inextricably linked. But π–π interactions provide stabilizing forces that outweigh the size issues even though these forces are exerted *on the outside* of the assembly. Folding, here and elsewhere, buries exposed surfaces and minimizes vacuums (surfaces that are not solvated). Folding inevitably liberates solvent and increases overall intramolecular interactions that stabilize assemblies.

Deconstruction of Capsules

Imperfect Walls

Given the difficulty in functionalizing concave surfaces, it seemed unlikely that the interior of the capsules could be outfitted with, say, catalytically active or other reagents to perform chemistry on resident guests. Besides, there is not enough room. But access of these agents from the outside was obvious, provided that windows could be built into the walls. This notion had additional appeal in that it offered a different paradigm for selectivity: namely, only that part of the resident guest exposed by the window could react with the agent outside. Wei Jiang set out to implement this approach by building a capsule with imperfect walls or windows to the outside.

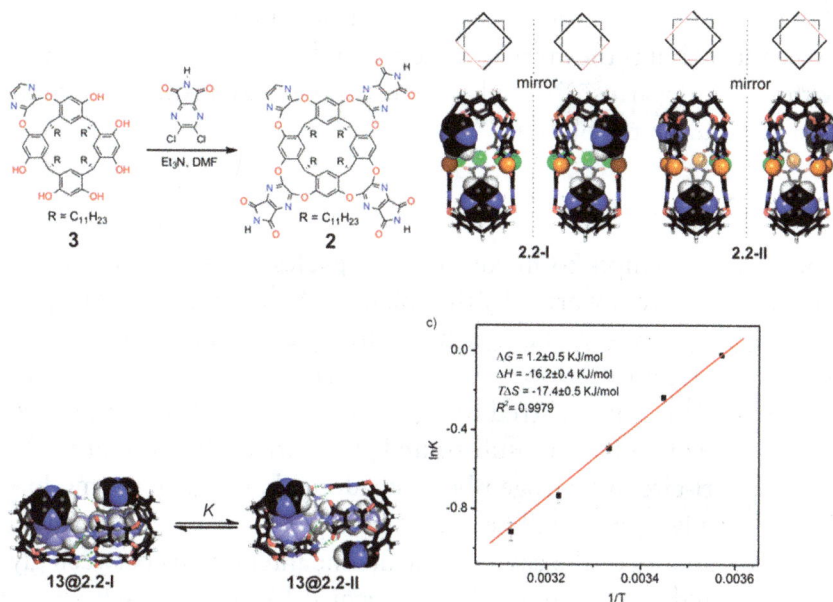

Figure 7.23 *Top*: Synthesis of a cavitand with one short wall and its (racemic) dimeric capsules. *Bottom*: Unlike many other guests that select between isomeric capsules, *n*-octylbenzene occupies both capsule isomers; the van 't Hoff plot for this guest is also shown. [Reprinted with permission from *J Am Chem Soc* **134**, 17498–17501. Copyright 2012, American Chemical Society.]

The monofunctionalized resorcinarene (Figure 7.23) was deepened with the usual walls to give the cavitand with a single plane of symmetry. It dimerizes to give two isomeric capsules (and their mirror images). In one isomer the short walls are proximal, giving an assembly with a sizable window, one that extends over both halves of the capsule.[88] In the other, the short walls are distal, giving two smaller windows, one in each half of the capsule.

A number of known guests of the fully intact cylindrical capsule (Chapter 3) were screened. They were not only encapsulated — they showed seemingly exaggerated preferences for one host over the other. Even though a guest might see little difference between the two isomers, the close quarters make for considerable and unpredictable selectivity. The larger window allows the approach of a reagent in solution to the central part of an encapsulated guest while the ends of the guest remain protected. This application remains for development in the future but is the inverse of the possibilities offered by complexes with "dangling arms"[89] — where parts of the guest protrude from the host structure into solution.[90]

Deep Cavitands

The earlier attempts to make longer capsules by dimerizing deeper cavitands[91] were thwarted by the collapse of the larger aromatic panels onto themselves, a phenomenon initially encountered by Cram.[92] The dimerization that occurred led to structures that had no internal spaces.[93] But it occurred to Dariush Ajami and Yoshi Yamauchi that the use of an N-monosubstituted glycoluril such as (Figure 7.24) would short-circuit hydrogen-bonding along that edge of the module. The other edge would still be available to pair its best acceptors with the imide donors of the cavitand, while the interactions between glycolurils could remain intact. It also seemed likely that a self-sorting process would apply during the assembly and each cavitand should yield a single enantiomer of the deepened container structure.

We prepared the appropriately soluble N-methylated glycoluril by monomethylating **7.4c** (see Figure 7.2). We then used normal

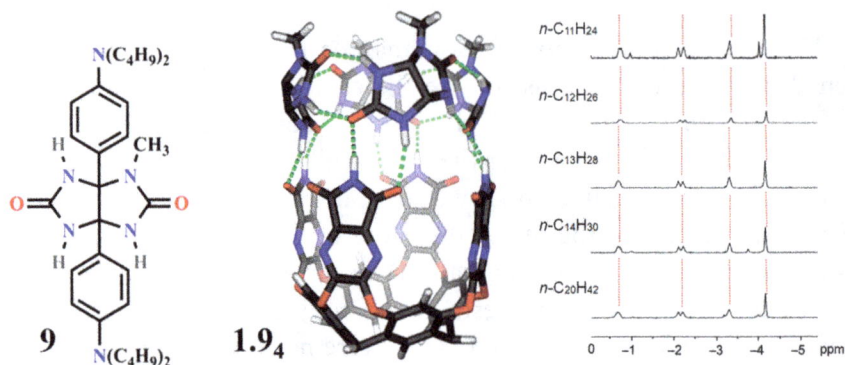

Figure 7.24 *Left*: Structure of mono-N-methylated glycoluril, and a model of its proposed assembly into a deepened cavitand. Peripheral groups have been deleted. *Right*: Partial NMR spectra of alkanes in the deepened cavitand. [Reprinted with permission from *Angew Chem Int Ed Eng* 50(39), 9150–9153. Copyright 2011, Wiley-VCH, Weinheim.]

alkanes encapsulated in the original dimeric cylinder and added the N-methylated spacer. The new, deepened cavitand host (Figure 7.24) emerged with alkanes partially inside and dangling outside.[94] Alkanes longer than n-C11 (which is an ideal guest for the original capsule) gave exclusively the new, open-ended assemblies. This was obvious from their NMR spectra (Figure 7.24); all the alkanes have the same signals for their first four carbons, and there is no gradual upfield shift that results from coiling of the longer alkanes in a closed (capsule) container. The diastereotopic splitting of the CH_2 signals indicates a chiral magnetic environment. These guests all exist[95] in an extended conformation; they experience the same aromatic envelope and apparently they go in and out the same way.[96]

We resolved the N-methyl glycoluril on a chiral column and repeated the experiment with the optically active material. Identical spectra were obtained: all four imide N–H's are equivalent and appeared at ~13.5 ppm, indicating a C_4 symmetry for the assembly. In other words, each assembly involves self-sorting that leads to a single enantiomer of the glycoluril. Characteristic CD spectra were obtained for the optically active assemblies with n-C13 and n-C17 guests. The proposed complex structure is also supported by DOSY and 2D NOESY experiments. Alkanes are generally not good guests for open-ended cavitands in organic solvents, but work well in aqueous media.[97] The case at hand indicates that the glycolurils provide a favorable environment for normal alkanes.

Reconstruction of Capsules

Unexpectedly, the use of higher alkanes (n-C15 to n-C19) gave a new type of extended capsules. The NMR spectra indicate that the longer hydrocarbon chains begin to coil in the new but still chiral environment (Figure 7.25). The stoichiometry comprises four N-methyl glycourils and two cavitands for one guest, but two types of imide N–H signals indicate a reduced symmetry. The NOESY spectrum of this assembly indicates a capsule formed by bringing together two deepened cavitands with the extrusion of four glycolurils. This leaves only

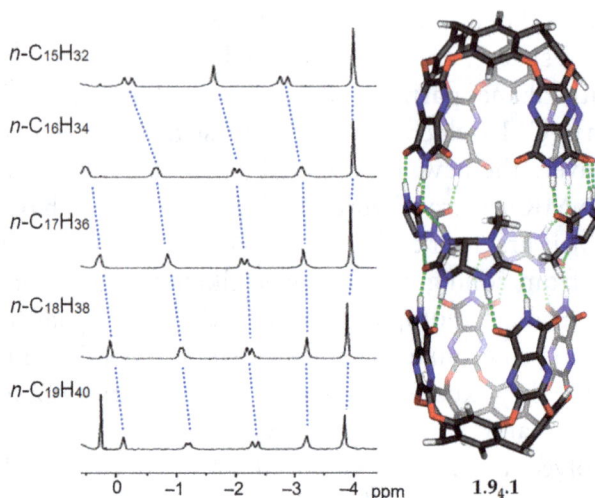

Figure 7.25 *Left*: Partial NMR spectra (600 MHz, 300 K, mesitylene-d_{12}) of complexes with specified normal alkanes. *Right*: Proposed model of a new capsule (the peripheral groups have been removed for clarity). [Reprinted with permission from *Angew Chem Int Ed Eng* 50(39), 9150–9153. Copyright 2011, Wiley-VCH, Weinheim.]

four hydrogen bonds holding the two halves of the assembly together. The proposed structure is supported by DOSY experiments.

The N-methyl groups of the glycolurils prevent the incorporation of more belts of spacers, so longer alkanes such as *n*-C20 are not encapsulated. Instead, they behave as the shorter alkanes and induce the formation of the deepened, chiral cavitand. A deepened cavitand on either end of the long alkane chain giving a dumbbell-shaped assembly is also possible, although such an assembly would involve bringing together nine molecules. But the lessons of Chapter 2 show that even more entropy-defying capsules are possible.

References

1. (a) Zanotti G, Scapin G, Spadon P, *et al.* (1992) Three-dimensional structure of recombinant human muscle fatty acid-binding protein. *J Biol Chem* 267(26), 18541–18550; (b) Han GW, Lee JY, Song HK, *et al.* (2001) Structural basis of non-specific lipid binding in maize lipid-transfer protein complexes revealed by high-resolution X-ray crystallography. *J Mol Biol* 308(2), 263–278.

2. Petina O, Rehder DE, Haupt TK, *et al.* (2011) Guests on different internal capsule sites exchange with each other and with the outside. *Angew Chem Intl Ed* **50**, 410–414.

3. Ko YH, Kim Y, Kim H, Kim K. (2011) U-shaped conformation of alkyl chains bound to a synthetic receptor cucurbit[8]uril. *Chem Asian J* **6**, 652–657.

4. Cavarzan A, Scarso A, Sgarbossa p, *et al.* (2011) Supramolecular control on chemo- and regioselectivity via encapsulation of (NHC)-Au catalyst within a hexameric self-assembled host. *J Am Chem Soc* **133**, 2848–2851.

5. Dydio P, Rubay C, Gadzikwa T, *et al.* (2011) Cofactor-controlled enantioselective catalysis. *J Am Chem Soc* **133**, 17176–17179.

6. Tucci FC, Rudkevich DM, Rebek Jr J. (1999) Deeper cavitands. *J Org Chem* **64**, 4555–4559.

7. Tucci FC, Rudkevich DM, Rebek Jr J. (2000) Velcrands with snaps and their conformational control. *Chem Eur J* **6**, 1007–1016.

8. Ajami D, Rebek Jr J. (2006) Expanded capsules with reversibly added spacers. *J Am Chem Soc* **128**, 5314–5315.

9. Rivera JM, Craig SL, Martín T, Rebek Jr J. (2000) Chiral guests and their ghosts in reversibly-assembled hosts. *Angew Chem Int Ed Eng* **39**, 2130–2132.

10. Scarso A, Trembleau L, Rebek Jr J. (2003) Encapsulation induces helical folding of alkanes. *Angew Chem Intl Ed* **42**, 5499–5502.

11. Ajami D, Rebek Jr J. (2006) Coiled molecules in spring-loaded devices. *J Am Chem Soc* **128**, 15038–15039.

12. (a) Ajami D, Rebek Jr J. (2006) Extended capsules with reversibly added spacers. *J Am Chem Soc* **128**, 5314–5315; (b) Ajami D, Rebek Jr J. (2007) Adaptations of guest and host in expanded self-assembled capsules. *Proc Natl Acad Sci USA* **104**, 16000–16003.

13. Schleyer PvR, Maerker C, Dransfeld A, *et al.* (1996) Nucleus-independent chemical shifts: a simple and efficient aromaticity Probe. *J Am Chem Soc* **118**, 6317–6318.

14. Frisher MJ *et al.* Gaussian 03 (Gaussian, Inc, Pittsburgh, PA, 2002).

15. Ajami D, Rebek Jr J. (2009) Multicomponent, hydrogen-bonded cylindrical capsules. *J Org Chem* **74**, 6584–6591.

16. Lützen A, Renslo AR, Schalley CA, *et al.* (1999) Encapsulation of ion–molecule complexes: second-sphere supramolecular chemistry. *J Am Chem Soc* **121**, 7455–7456.

17. Ajami D, Rebek Jr J. (2009) Compressed alkenes in reversible encapsulation complexes. *Nat Chem* **1**, 87–90.

18. Ajami D, Rebek Jr J. (2008) Gas behavior in self-assembled capsules. *Angew Chem Intl Ed* **47**, 6059–6061.

19. Ajami D, Rebek Jr J. (2008) Reversible encapsulation of terminal alkenes and alkynes. *Heterocycles* **76**, 169–176.

20. Detering C, Tolstoy PM, Golubev NS, *et al.* (2001) Vicinal H/D isotope effects in NMR spectra of complexes with coupled hydrogen bonds. *Dok Phys Chem* **379**, 1–4.

21. Shenderovich IG, Limbach HH, Smirnov SN, *et al.* (2002) H/D isotope effects on the low-temperature NMR parameters and hydrogen bond geometries of $(FH)_2$ F-dissolved in CSF_3/CDF_2CL. *Phys Chem Chem Phys* **4**, 5488–5497.

22. Tolstoy PM, Schah-Mohammedi P, Smirnov SN, et al. (2004) Characterization of fluxional hydrogen-bonded complexes of acetic acid and acetate by NMR: geometeries and isotope and solvent effects. *J Am Chem Soc* **126**, 5621–5634.

23. Limbach HH, Denisov GS, Golubev NS. (2005) Hydrogen bond isotope effects studied by NMR. In: Kohen A, Limbach HH (eds.), *Isotope Effects in Chemistry and Biology*, Taylor & Francis, Boca Raton, FL, Chap. 7, pp. 193–230.

24. Harris RK, Jackson P, Merwin LK, et al. (1988) *J Chem Soc Faraday Trans I* **84**, 3649–3672.

25. (a) Sternberg U, Brunner EL. (1994) The influence of short-range geometry on the chemical shift of protons in hydrogen bonds. *J Magn Res A* **108**, 142–150; (b) Brunner E, Sternberg U. (1998) Solid-state NMR investigations on the nature of hydrogen bonds. *J Progr NMR Spect* **32**, 21–57.

26. Mildvan TK, Harris AS. (1999) *Proteins Struct Funct Genet* **35**, 275–282.

27. McDermott A, Ridenour CF. (1996) Proton chemical shift measurements in biological solids. In *Encyclopedia of NMR*. Wiley, Sussex, UK, pp. 3820–3825.

28. Emmler T, Gieschler S, Limbach HH, Buntkowsky G. (2004) *J Mol Struct* **700**, 29–38.

29. Ajami D, Tolstoy P, Dube H, et al. (2011) Encapsulated carboxylic dimers and compressed hydrogen bonds. *Angew Chem Int Ed* **50**, 528–531.

30. Shivanyuk A, Rebek Jr J. (2002) Social isomers in encapsulation complexes. *J Am Chem Soc* **124**, 12074–12075.

31. Avram L, Cohen Y. (2004) Self-recognition, structure, stability, and guest affinity of pyrogallol[4]arene and resorcin[4]arene capsules in solution. *J Am Chem Soc* **126**, 11556–11563.

32. Shivanyuk A, Rebek Jr J. (2001) Reversible encapsulation of multiple, neutral guests in hexameric resorcinarene hosts. *Chem Commun* 2424–2425.

33. Hof F, Nuckolls C, Rebek Jr J. (2000) Diversity and selection in self-assembled tetrameric capsules. *J Am Chem Soc* **122**, 4251–4252.

34. Kerckhoffs JMCA, ten Cate MGJ, Mateos-Timoneda MA, et al. (2005) Selective self-organization of guest molecules in self-assembled molecular boxes. *J Am Chem Soc* **127**, 12697–12708.

35. MacGillivary LR, Atwood JL. (1997) A chiral spherical molecular assembly held together by 60 hydrogen bonds. *Nature* **389**, 469–472.

36. Gerkensmeier T, Iwanek W, Agena C, et al. (1999) Self-assembly of 2,8,14,20-tetraisobutyl-5,11,17,23-tetrahydroxyresorcin[4]arene. *Eur J Org Chem* **9**, 2257–2262.

37. Shivanyuk A, Rebek Jr J. (2003) Assembly of resorcinarene capsules in wet solvents. *J Am Chem Soc* **125**, 3432–3433.

38. Kobayashi K, Ishii K, Sakamoto S, et al. (2003) Guest-induced assembly of tetracarboxyl-cavitand and tetra(3-pyridyl)-cavitand into a heterodimeric capsule via hydrogen bonds and CH-halogen and/or CH-π interaction: control of the orientation of the encapsulated guest. *J Am Chem Soc* **125**, 10615–10624.

39. Gonzalez JJ, Ferdani R, Albertini E, et al. (2000) Dimeric capsules by the self-assembly of triureidocalix[6]arenes through hydrogen bonds, *Chem Eur J* **6**, 73–80.

40. Scarso A, Pellizzaro L, De Lucchi O, *et al.* (2007) Gas hosting in enantiopure self-assembled oximes. *Angew Chem Int Ed* **46**, 4972–4975.
41. Sherman JC. (1995) Carceplexes and hemicarceplexes: Molecular encapsulation — from hours to forever. *Tetrahedron* **51**, 3395–3422.
42. Brody MS, Schalley CA, Rudkevich DM, Rebek Jr J. (1999) Synthesis and characterization of an intramolecularly self-assembled capsule. *Angew Chem Int Ed Eng* **38**, 1640–1644.
43. Yoshizawa M, Tamura M, Fujita M. (2006) Diels–Alder in aqueous molecular hosts: unusual regioselectivity and efficient catalysis. *Science* **312**, 251–254.
44. Ziegler M, Brumaghim JL, Raymond KN. (2000) Stabilization of a reactive cationic species by supramolecular encapsulation. *Angew Chem Int Ed* **39**, 4119–4121.
45. Kaanumalle LS, Gibb CLD, Gibb BC, Ramamurthy V. (2005) A hydrophobic nanocapsule controls the photophysics of aromatic molecules by suppressing their favored solution pathways. *J Am Chem Soc* **127**, 3674–3675.
46. Ajami D, Dube H, Rebek Jr J. (2011) Boronic acid hydrogen bonding in encapsulation complexes. *J Am Chem Soc* **133**, 9689–9691.
47. Nishimura N, Kobayashi K. (2008) Self-assembly of a cavitand-based capsule by dynamic boronic ester formation. *Angew Chem Int Ed* **47** 6255–6258.
48. Larkin JD, Bhat KL, Markham GD, *et al.* (2006) Structure of the boronic acid dimer and the relative stabilities of its conformers. *J Phys Chem A* **110**, 10633–10642.
49. Ajami, D, Rebek Jr J. (2007) Longer guests drive the reversible assembly of hyperextended capsules. *Angew Chem Intl Ed* **46**(48), 9283–9286.
50. Ajami D, Liu L, Rebek Jr J. (2014) Soft templates in encapsulation complexes. *Chem Soc Rev*, in press, DOI:10.1039/C4CS00065J.
51. Mecozzi S, Rebek Jr J. (1998) The 55% solution: a formula for molecular recognition in the liquid state. *Chem Eur J* **4**, 1016–1022.
52. Palmer LC, Rebek Jr J. (2004) The ins and outs of molecular encapsulation. *Org Biomol Chem* **2**, 3051–3059.
53. Whitesides GM, Mathias JP, Seto CT. (1991) Molecular self-assembly and nanochemistry: a chemical strategy for the synthesis of nanostructures. *Science*, **254**, 1312.
54. Rebek Jr J. (1979) Mechanistic studies using solid supports: the three-phase test. *Tetrahedron* **35**, 723–732.
55. Tjivikua T, Ballester P, Rebek Jr J. (1990) A self-replicating system. *J Am Chem Soc* **112**, 1249–1250.
56. Devane WA, Hanus L, Breuer A, *et al.* (1992) Isolation and structure of a brain constituent that binds to the cannabinoid receptor. *Science* **258**, 1946–1949.
57. Ajami D, Rebek Jr J. (2007) Adaptations of guest and host in expanded self-assembled capsules. *Proc Natl Acad Sci USA* **104**, 16000–16003.
58. Cravatt BF, Lerner RA, Boger DL. (1996) Structure determination of an endogenous sleep-inducing lipid, cis-9-octadecenamide (oleamide): a synthetic approach to the chemical analysis of trace quantities of a natural product. *J Am Chem Soc* **118**, 580–590.

59. Taira T, Ajami D, Rebek Jr J. (2012) Encapsulation of ion pairs in extended, self-assembled structures. *J Am Chem Soc*, **134**, 11971–11973.
60. Dega-Szafran Z, Grundwald-Wyspianska M, Szafran, M. (1991) *Spectrochim Acta Part A* **47**, 543–550.
61. Taira T, Ajami D, Rebek Jr J. (2012) Hydration of isocyanates in an expandable, self-assembled capsule. *Chem Commun* **48**, 8508–8510.
62. Balzani V, Credi A. Venturi M, eds. (2003) *Molecular Devices and Machines: A Journey into the Nanoworld*. Wiley-VCH, pp. 288–328.
63. Kay ER, Leigh DA, Zerbetto F. (2007) Synthetic molecular motors and mechanical machines. *Angew Chem Int Ed* **46**, 72–191.
64. Deal JR, WJ, Erlanger BF, Nachmansohn D. (1969) Photoregulation of biological activity by photochromic reagents. III. Photoregulation of bioelectricity by acetylcholine receptor inhibitors. *Proc Natl Acad Sci USA* **64**, 1230–1234.
65. Banghart MR, Mourot A, Forti, DL, *et al.* (2009) Photochromic blockers of voltage-gated potassium channels. *Angew Chem Int Ed* **48**, 9097–9101.
66. Shinkai S, Ogawa T, Nakaji T, *et al.* (1979) Photocontrolled extraction ability of azobenzene-bridged azacrown ether. *Tetrahedron Lett* **20**, 4569–4572.
67. Shinkai S, Nakaji T, Nishida Y, *et al.* (1980) Photoresponsive crown ethers. 1. Cis–trans isomerism of azo-benzene as a tool to enforce conformational changes of crown ethers and polymers. *J Am Chem Soc* **102**, 5860–5865.
68. Shinkai S, Nakaji T, Ogawa T, *et al.* (1981) Photoresponsive crown ethers. 2. Photocontrol of ion extraction and ion transport by a bis(crown ether) with a butterfly-like motion. *J Am Chem Soc* **103**, 111–115.
69. Gloe K, ed. (2005) *Macrocyclic Chemistry: Current Trends and Future Perspectives*. Springer, The Netherlands, pp. 203–218.
70. Wang Y, Ma N, Wang Z, Zhang X. (2007) Photocontrolled reversible supramolecular assemblies of an azobenzene-containing surfactant with a-cyclodextrin. *Angew Chem Int Ed* **46**, 2823–2826.
71. Dabrowski R, Kenig K, Raszewski Z, *et al.* (1980) Synthesis and some physical properties of unsymmetrical 4,4'-dialkylazoxybenzenes. *Mol Cryst Liq Cryst* **61**, 61–78.
72. Dube H, Ajami D, Rebek Jr J. (2010) Photochemical control of reversible encapsulation. *Angew Chem Int Ed* **49**, 3192–3195.
73. Dube H, Rebek Jr J. (2012) Selective guest exchange in encapsulation complexes using different light inputs. *Angew Chem Int Ed* **51**, 3207–3210.
74. Moon K, Chen WZ, Ren T, Kaifer AE. (2003) A unique hydrogen bonding network in the crystal structure of 3a,6a-diphenylglycoluril. *Crystengcomm* **5**, 451–453.
75. Piacenza G, Beguet C, Wimmer E, *et al.* (1997) 2,4,6,8-tetraazabicyclo-[3.3.1]nonane-3,7-dione and 2,4,6,8-tetraacetyl-2,4,6,8-tetraazabicyclo[3.3.1]-nonane-3,7-dione. *Acta Crystallogr C* **53**, 1459–1462.
76. Tiefenbacher K, Ajami D, Rebek Jr J. (2011) Self-assembled capsules of unprecedented shapes. *Angew Chem Int Ed* **50**, 12003–12007.
77. Siering C, Torang J, Kruse H, *et al.* (2010) Enantioselective helical folding inside a self-assembled, cylindrical capsule. *Chem Commun* **46**(10), 1625–1627.

78. Ajami D, Rebek Jr J. (2009) Compressed alkanes in reversible encapsulation complexes. *Nat Chem* 1(1), 87–90.

79. Kaanumalle LS, Gibb CLD, Gibb BC, Ramamurthy V. (2005) A hydrophobic nanocapsule controls the photophysics of aromatic molecules by suppressing their favored solution pathways. *J Am Chem Soc* 127, 3674.

80. (a) Burley SK, Petsko GA. (1985) Aromatic–aromatic interaction — a mechanism of protein-structure stabilization. *Science* 229, 23–28; (b) Blundell TL, Singh J, Thornton J, *et al.* (1986) Aromatic interactions. *Science* 234, 1005; (c) Galán A, de Mendoza J, Toiron C, *et al.* (1991) A synthetic receptor for dinucleotides. *J Am Chem Soc* 113, 9424–9425; (d) Jeong KS, Muehldorf AV, Rebek Jr J. (1990) Molecular recognition. Asymmetric complexation of diketopiperazines. *J Am Chem Soc* 112, 6144–6145.

81. Paliwal S, Geib S, Wilcox CS. (1994) Molecular torsion balance for weak molecular recognition forces. Effects of "tilted-T" edge-to-face aromatic interactions on conformational selection and solid-state structure. *J Am Chem Soc* 116, 4497–4498.

82. Hof F, Scofield DM, Schweizer WB, Diederich F. (2004) A weak attractive interaction between organic fluorine and an amide group. *Angew Chem Int Ed* 43, 5056–5059.

83. Mati IK, Cockroft SL. (2010) Molecular balances for quantifying non-covalent interactions. *Chem Soc Rev* 39, 4195–4205.

84. Tiefenbacher K, Rebek Jr J. (2012) Selective stabilization of self-assembled hydrogen-bonded molecular capsules through π–π interactions. *J Am Chem Soc* 134, 2914–2917.

85. Castellano RK, Meyer EA, Diederich F. (2003) Interactions with aromatic rings in chemical and biological recognition. *Angew Chem Int Ed* 42, 1210–1250.

86. Salonen LM, Ellermann M, Diederich F. (2011) Aromatic rings in chemical and biological recognition: energetics and structures. *Angew Chem Int Ed* 50, 4808–4842.

87. Bissantz C, Kuhn B, Stahl M. (2010) A medicinal chemist's guide to molecular interactions. *J Med Chem* 53, 5061–5084.

88. Jiang W, Rebek Jr J. (2012) Guest-induced, selective formation of isomeric capsules with imperfect walls. *J Am Chem Soc* 134, 17498–17501.

89. Tiedemann BEF, Raymond KN. (2006) Dangling arms: a tetrahedral supramolecular host with partially encapsulated guests. *Angew Chem Int Ed* 45, 83–86.

90. Kobayashi K, Yamanaka M. (2014) Self-assembled capsules based on tetra-functionalized clix[4]resorcinarene cavitands. *Chem Soc Rev*, DOI:10.1039/c4cs00153b.

91. Tucci FC, Rudkevich DM, Rebek Jr J. (1999) Deeper cavitands. *J Org Chem* 64, 4555–4559.

92. Cram DJ, Choi HJ, Bryant JA, Knobler CB. (1992) Host–guest complexation. 62. Solvophobic and entropic driving forces for forming velcraplexes, which are 4-fold, lock–key dimers in organic media. *J Am Chem Soc* 114(20), 7748–7765.

93. Tucci FC, Rudkevich DM, Rebek Jr J. (2000) Velcrands with snaps and their conformational control. *Chem Eur J* 6(6), 1007–1016.
94. Yamauchi Y, Ajami D, Lee JY, Rebek Jr J. (2011) Deconstruction of capsules using chiral spacers *Angew Chem Int Ed* 50, 9150–9153.
95. Trembleau L, Rebek Jr J. (2003) Helical conformation of alkanes in hydrophobic environments. *Science* 301, 1219–1220.
96. Rieth S, Hermann K, Wang BY, Badjic JD. (2011) Controlling the dynamics of molecular encapsulation and gating. *Chem Soc Rev* 40, 1609–1622.
97. Scarso A, Trembleau L, Rebek Jr J. (2003) Encapsulation induces helical folding of alkanes. *Angew Chem Intl Ed* 42(44), 5499–5502.

CHAPTER 8

Reactions Inside Capsules

Introduction

What practical applications are available to these hydrogen-bonded capsules? Almost everyone proposes drug delivery. To be sure, many medicines fail in clinical trials because of their physical properties; they are incompatible with normal physiology, despite the efforts of the world's best medicinal chemists and formulators. But consider the cost of, say, a molecule of taxol versus the cost of one of these capsules. If the delivery is stoichiometric, then the container costs much more than the payload. An extreme case (proposed and actually started by a US oil company) was the mining of methane from its mountainous clathrates in the oceans. Besides, most of the hydrogen-bonded capsules lack solubility in biological conditions. An exception might be in transport of medicines across membrane barriers. This could be practical if the capsules could act as shuttles and turn over a large number of times. The ideal payload would be a medicine that acts as an efficient *catalyst*, rather than a stoichiometric agent. In the absence of these applications, we looked for uses of the capsules as reaction flasks.

There are many ways to accelerate reactions, including increases in reagent concentration, increases in temperature, and catalysis by functional groups that interact with and lower the energy of the transition state. With compartmentalized reagents, there is another dimension, namely, time. Molecules that are coencapsulated encounter each other for millions of times longer than they do in bulk solution. In solution molecules diffuse into the same solvent cage and typically interact for only a few nanoseconds before the cage dissipates and the molecules drift apart. Capsules *are* solvent cages, and the synthetic efforts that have gone into their synthesis result in robustness: the capsule's panels that surround the reactants are not free to depart or even move much. In some ways, they represent the ultimate in solvent cages. One might think that the panels, because of their immobility, would not favor formation of a transition state from an encapsulated ground state, but there has been no evidence for such a hindering effect. Perhaps the "breathing motions" of hydrogen-bonded capsules provide enough flexibility in the panels to adjust favorably to transition states as they form inside.

There is one seemingly insurmountable disadvantage. Functional groups in the lining of the capsule that could be brought to bear on the reactants and act as chemical catalysts — acids or bases — would be ideal but are unlikely to be available soon. The reason lies in the difficulty of functionalizing concave surfaces. Enzymes solved this problem by starting with a linear structure which folds around the substrates in a way that presents the reaction to be catalyzed with the appropriate amino acid side chains. This result is through the wisdom of evolution over billions of years. For chemists, the timescale of a few decades does not inspire confidence that the catalysis problem will be solved. While there are acids and bases imbedded in the seam of hydrogen bonds, these are generally orthogonal to, rather than convergent on the reagents held inside, and are unlikely to provide effective hydrogen-bonding to transition states. They might do so through bifurcated hydrogen bonds but this would be simply lucky.

However, capsules do offer some advantages. First, consider the enhanced reactant concentrations: these are easily calculated on the basis of a capsule's volume and how many molecules occupy it. These are calculations of *real* concentrations, rather than effective molarities, and they are not matters of distance. Molecules confined in a small space for prolonged periods of time will reach an arrangement conducive for reaction, if they are not prevented from doing so. The capsule volumes of 300–1200 Å^3 translate to $3 - 12 \times 10^{-25}$ L, and 1 molecule is 16.6×10^{-25} moles. Therefore, concentrations are typically about 1–5 M. Any molecule inside these capsules enjoys at least *single digit molar* concentrations.

The Diels–Alder Reaction

The softball binds two molecules of benzene (~9 M!) and the roughly spherical shape of the space in the softball requires the two aromatic guests to be face-to-face if they are to fit inside. The ready encapsulation of compounds that resemble two face-to-face aromatic molecules, such as paracyclophane and ferrocene, is another indication of this rule. This configuration resembles the transition structure of a typical Diels–Alder reaction involving maximum overlap of unsaturated centers, and it was inevitable that Jongmin Kang would try this reaction first in the softball.[1] The Diels–Alder reaction between *p*-quinone and cyclohexadiene in typical organic solvents shows a half-life of two days at molar concentrations of the reaction partners (Figure 8.1). At the millimolar concentrations of typical NMR experiments, this reaction would have a half-life on the order of years. On exposure of cyclohexadiene and *p*-quinone to the capsule in *p*-xylene-d_{10}, the resting state of the capsule shows a symmetrical structure in the NMR spectrum in which two quinones are present. At these concentrations cyclohexadiene cannot compete with solvent for the capsule. At concentrations of cyclohexadiene >0.2 M it can compete enough to allow its observation by NMR and permit calculation of the apparent encapsulation constant as 22 M^{-2}.[2] Since capsules are never empty, the absolute values are unavailable, and this represents an apparent value for the tetramolecular complex (two molecules each of module and

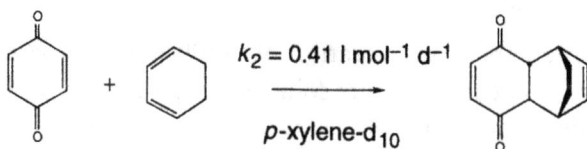

$k_2 = 0.41\ l\ mol^{-1}\ d^{-1}$

p-xylene-d_{10}

Figure 8.1   A conventional Diels–Alder reaction.

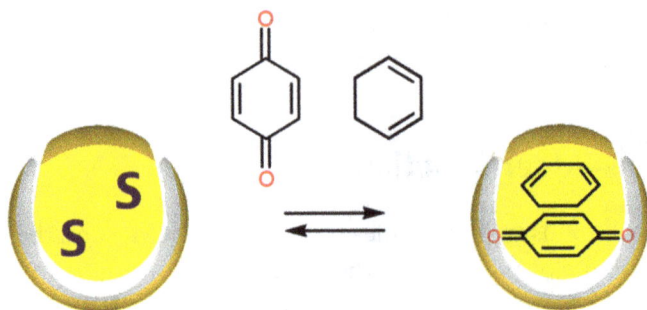

Figure 8.2   The resting state of the capsule contains two quinone guests, and occasionally one is replaced by a cyclohexadiene molecule to form the "Michaelis" complex.

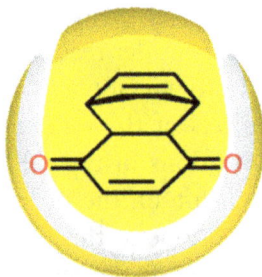

Figure 8.3   The Diels–Alder adduct gradually fills the capsule and the encapsulated reaction ends in a classic version of product inhibition.

guest). The corresponding value for p-quinone is $1.9 \times 10^5 \, \mathrm{M}^{-2}$ and neither capsules with two molecules of cyclohexadiene, nor those with one of each reactant, are apparent in the NMR spectra (Figure 8.2). Even so, a small concentration of the latter must be present, since within a few hours the Diels–Alder adduct begins to appear in the capsule (Figure 8.3).

This represents an acceleration of ~200-fold compared to what is going on outside in the dilute solution. Experiments with increased cyclohexadiene concentrations showed saturation kinetics (Figure 8.4) and established that the reaction takes place in the capsule.

At first glance, this is an impressive acceleration, but on second glance the concentration of the two components, roughly 4 M

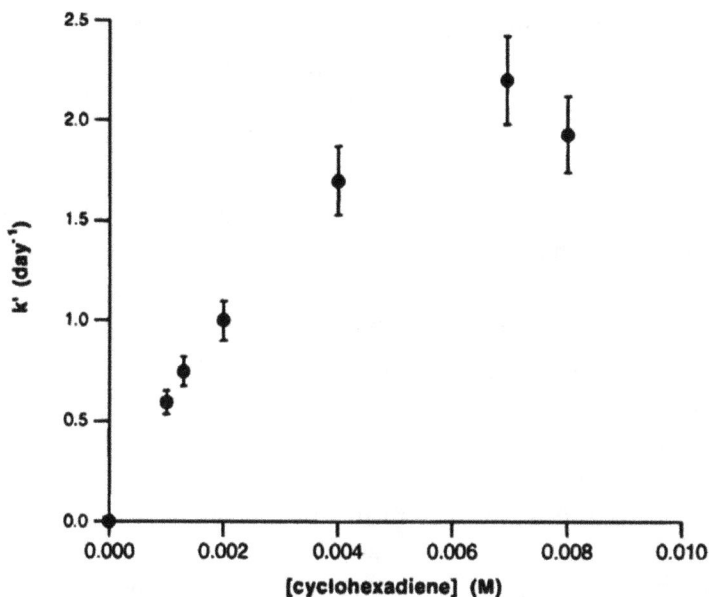

Figure 8.4   Pseudo first order rate constants of the reaction of maleic anhydride with cyclohexadiene as a function of the latter's concentration. [Reprinted with permission from *J Am Chem Soc* **120**, 7389–7390. Copyright 1998, American Chemical Society.]

or 5 M apiece, shows that the acceleration is nothing special. That is to say, there is no chemical catalysis involved. Instead, the "pseudoconcentration" of the two reagents when they both occupy the capsule is the cause of the rate enhancement.

Eventually, the capsule is filled with the Diels–Adler reaction product, it is the single (endo) isomeric product that involves the maximum overlap of unsaturated centers. But the product — one molecule, in contrast to the two of the starting materials — is entropically favored inside. The adduct gradually fills the capsule and the reaction grinds to a halt in a classic version of product inhibition. The Diels–Alder adduct is a superb guest and its binding affinity is too large to determine accurately by NMR titrations. Through competition experiments, an estimate for the apparent association constant ($K_a \sim 10^4$ M^{-1}) was made.

Now, product inhibition is a phrase coined by biochemists — a natural phenomenon, one that is an inevitable consequence of reactions that feature transition states that closely resemble the products. This is one reason why there are very few "Diels–Alderases" in Nature. In other words, turnover is thwarted by the entropically favored, single-molecule product over the encapsulation of two molecular reactants.

Further indication that the Diels–Alder reaction takes place inside the capsule was provided by competition experiments with good, single-molecule guests. For example, paracyclophane effectively shut down the reaction, as did the addition of the adduct itself. Although there was product inhibition by the capsules, there was also an exquisite selectivity: molecules that do not fit inside, such as naphthoquinone, are not accelerated in their reaction with cyclohexadiene. The acceleration by hydrogen bonding to the quinone *inside* the confined environment of the capsule could not be ruled out, but controls with noncapsule molecules with the same functional groups showed no effect on the cycloaddition reaction rate. Specifically, the *S*-shaped isomer of the module (Figure 8.5), which cannot assemble into a capsule, was used in these control experiments and showed no acceleration.

The saturation kinetics, size selectivity and inhibition by other guests of the capsule all point to a reaction taking place within the capsule.

Other reaction pairs were also tested in this capsule (Figure 8.6). Of these, maleic anhydride showed comparable reactivity with cyclohexadiene. Unlike *p*-quinone, maleic anhydride exchanges rapidly

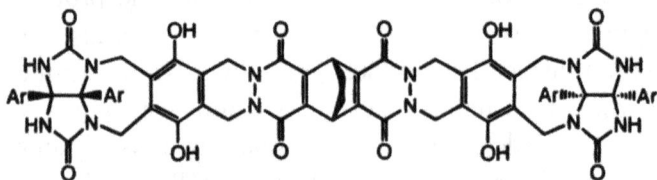

Figure 8.5 The *S*-shaped isomer of the softball module cannot assemble into a capsule, and showed no acceleration of the Diels–Alder reaction.

Figure 8.6 Other dienes and dienophiles tested in the softball.

in and out of the capsule, but its encapsulation constant could be determined to be 2×10^5 M^{-2} for the tetramolecular complex, or about the same as that of p-quinone. The Diels–Alder reactions of cyclopentadiene were not accelerated in the presence of the capsule. Presumably, this diene is too small to appropriately fill the space.

These results come as no surprise to enzymologists, who know that the active sites of enzymes are not just about rate enhancements, but also about selectivity. Enzymes are needed to regulate concentrations. In an effort to overcome product inhibition in the softball, Jongmin Kang, Javier Santamaria and Göran Hilmersson considered the reaction of 2,5-dimethyl-thiophenedioxide with p-benzoquinone. In solution, the initial adduct loses SO_2 and flattens the product into a naphthalene-like structure that does not fit well inside the softball. In the experiment, the reaction did show turnover — authentic catalysis — inside the softball, but not for the reasons that we had expected. The initial adduct was found to be replaced by more starting materials (quinones) in the capsule (Figure 8.7).[3] Apparently, the adduct is an uncomfortable fit and escapes the confines. A similar case was encountered by Fujita with Diels–Alder reactions in metal–ligand cavitands.[4] A dividend from this research was the actual isolation and characterization of the initial adduct, which has proven difficult to achieve in dilute solutions.

Some 15 years later the Diels–Alder reaction was performed in the hexameric capsule (Figure 8.8) by Shimizu *et al.*[5] The capsule was a special one, involving fluorous "feet,"[6] and a number of Diels–Alder partners (Figure 8.9) were successfully reacted in this capsule when control reactions showed low yields. These observations augur well for the application of the fluorophobic effect to chemistry in small spaces.[7]

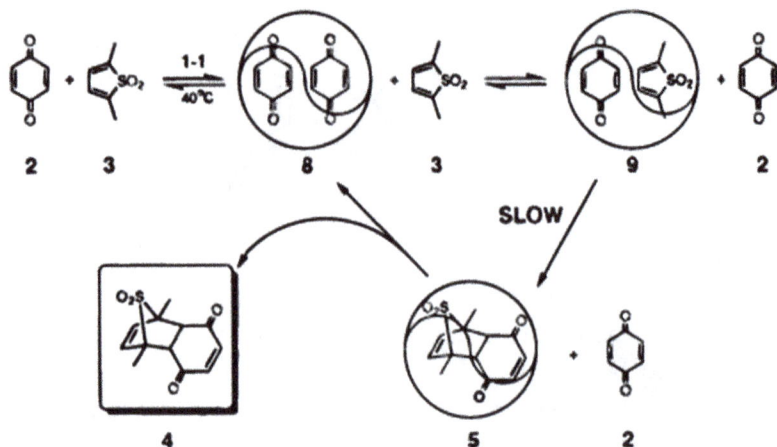

Figure 8.7   True catalysis by the softball: the quinone is a better guest and releases the adduct, forcing turnover.

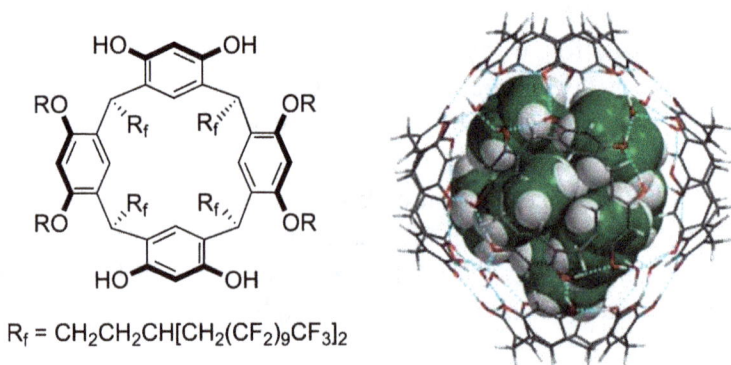

$R_f = CH_2CH_2CH[CH_2(CF_2)_9CF_3]_2$

Figure 8.8   *Left*: A resorcinarene with fluorous "feet." *Right*: Eight benzenes occupy the cavity in its resting state.

The "Click" Reaction

The second encapsulated reaction was the celebrated "click" cycloaddition reaction developed by Sharpless.[8] In organic solvents, the reaction at molar concentration takes place on the timescale of hours, but at millimolar concentrations the rate is diminishingly small, with half-lives of years, and the two regioisomers are

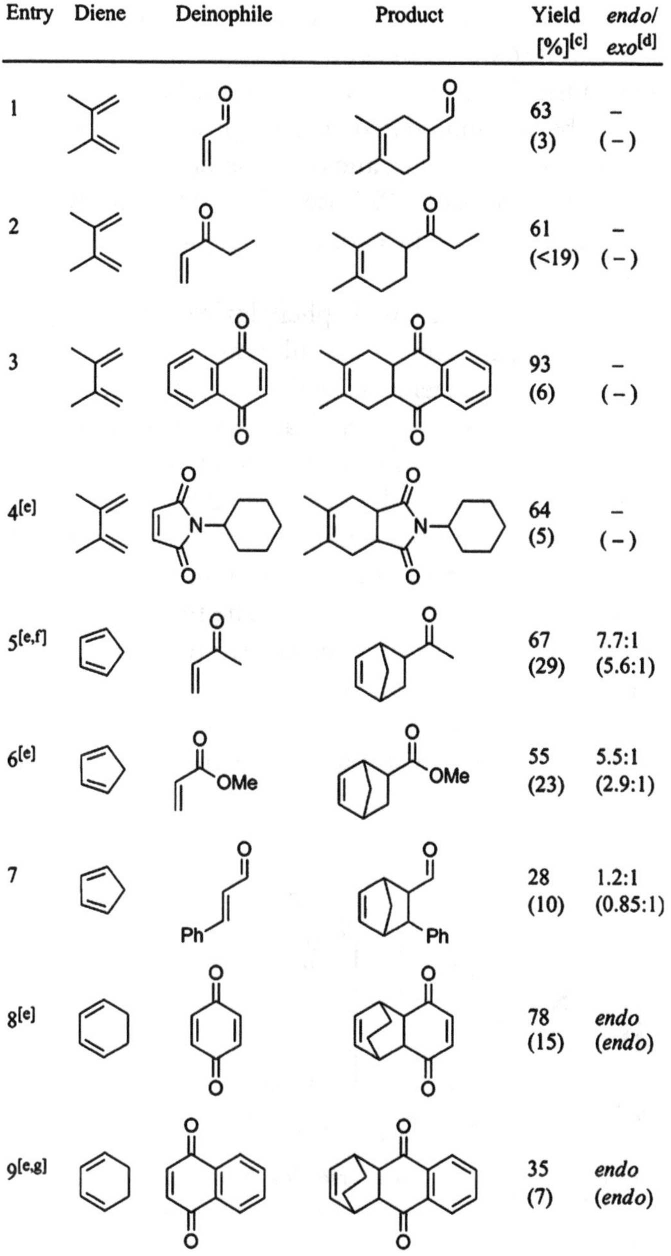

Entry	Diene	Deinophile	Product	Yield [%][c]	endo/ exo[d]
1				63 (3)	– (–)
2				61 (<19)	– (–)
3				93 (6)	– (–)
4[e]				64 (5)	– (–)
5[e,f]				67 (29)	7.7:1 (5.6:1)
6[e]				55 (23)	5.5:1 (2.9:1)
7				28 (10)	1.2:1 (0.85:1)
8[e]				78 (15)	endo (endo)
9[e,g]				35 (7)	endo (endo)

Figure 8.9 Diels–Alder reactions accelerated by a hexameric resorcinarene with fluorous feet. [Reprinted with permission from *Eur J Org Chem* 4734–4737. Copyright 2013, WILEY-VCH, Weinheim.]

formed in comparable amounts. Catalysis with copper ions devised by Sharpless and Fokin[9] tremendously accelerates this reaction. We used the cylindrical capsule, in which two different aromatic guest molecules can be accommodated. The shape of the space forces their orientation to be edge-to-edge, and only contact between their peripheral substituents is possible. We used this feature to evaluate intermolecular forces in the context of single solute–solvent interactions (Chapter 3).

The capsule was tested with phenylazide and phenylacetylene. Happily, on mixing, an isomeric capsule was observed with more than 80% involving the arrangement ideal for the cycloaddition reaction (Figure 8.10).[10] Both acetylene and azide were positioned near the center of the capsule, poised for the reaction to occur, and it did, giving a single regio-isomer accelerated some 20,000-fold compared to the reaction in dilute solution. Again, this was acceleration; there was no turnover, as that isomer of the cycloaddition product was the best of all possible guests inside the capsule. The reaction comes to a halt when the capsule is filled with the product. Treatment with methanol released a single stereoisomer. Neither molecules such as naphthylazide (which cannot fit inside this capsule) nor biphenylazide (which fits but leaves no room for a reaction partner) were accelerated in

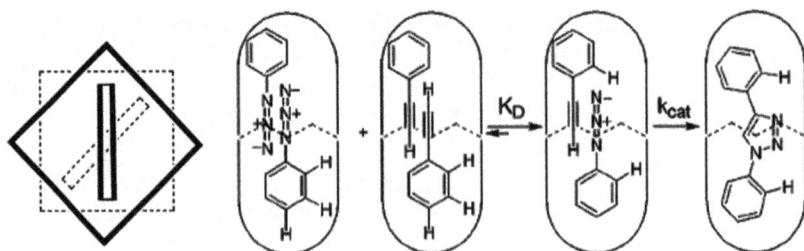

Figure 8.10 An encapsulated "click" reaction between phenylazide and phenylacetylene. On mixing, the major capsule observed contains one of each reagent, oriented for reaction.

their reactions. So, once again, there was exquisite selectivity, but no catalytic turnover.

We were fortunate to find the correct positioning of the primary acetylene, since this narrowest of functional groups can access the tapered ends of the capsule. Perhaps the remaining 20% of the capsules contain this social isomer, which equilibrates with the desired orientation during the course of the reaction. Regrettably, such exquisite selectivity is not valued in chemical synthesis, where generality in substrate scope is desired. Selectivity is, however, the rule in enzymatic reactions, where *regulation* is the function.

Chemical Amplification

The foregoing reaction accelerations were interpreted through the time and space which reactants enjoy in the capsule. Jian Chen, Steffi Körner, Stephen Craig, Shirley Lin and Dmitry Rudkevich joined forces to explore the consequence of the mechanical barrier that the capsules place between reagents inside and outside. Inside, a reagent like dicyclohexylcarbodiimide (DCC) is isolated and its reactivity as a dehydrating agent is turned off. Outside the capsule, the DCC is free to react with carboxylic acids to give dicyclohexyl urea (DCU) and anhydrides, active esters or amides, depending on what acyl acceptors are present in solution. Now DCC is an excellent guest for the capsule as it has a congruent shape, appropriate size and complementarity of chemical surfaces. But DCU is even better, as it has all of these attributes and can present both hydrogen bond donors and acceptors to interact with the capsule's polar seam. Accordingly, DCU rapidly displaces DCC from the capsule. This has a curious consequence for reactions that produce DCU: any process that forms DCU in solution will release DCC that is present in the capsule. The process is one of DCU self-replication but is not based on the familiar template effects or direct contact between DCC and DCU. [11] Instead, it results from exchange of compartments, and the release of DCC into solution fuels its conversion to DCU. With appropriate reagents, a true chain reaction could be engineered.

Chain Reaction Kinetics

We studied the reaction of *p*-toluic acid and *p*-ethylaniline in solution with encapsulated DCC. The anilide product is also an excellent guest and is capable of displacing DCC. The observed reaction profile (Figure 8.11) is distinctly sigmoidal and the capsule gradually fills with

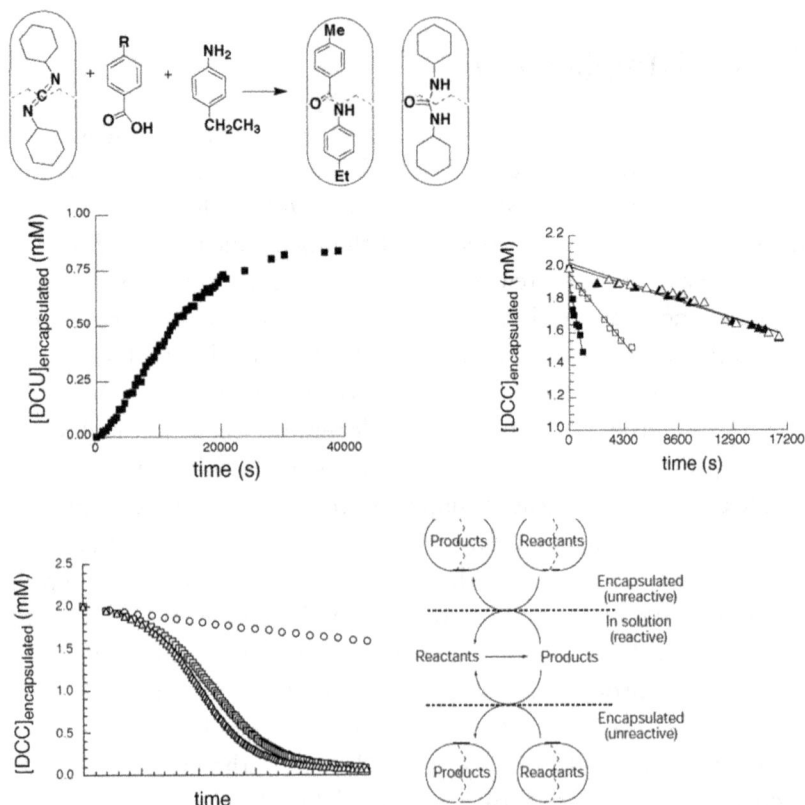

Figure 8.11   *Top*: Initial states of reagents and their chain reaction. *Center*: Product generation curves show sigmoidal behavior. *Bottom*: Simulated rates of DCC release with one, two or three products that displace DCC. [From *Proc Natl Acad Sci USA* **99**, 2593–2596. Copyright 2002, National Academy of Sciences of the United States of America.]

DCU and the anilide product.[12] The DCC is encapsulated but there are trace amounts of it in solution (its equilibrium concentration) at the beginning of the reaction. But as a DCC reacts to give anilide and urea, these two products displace two equivalents of DCC from the capsule and accelerate the reaction.

A qualitatively different behavior is shown by the longer *p*-ethylbenzoic acid. Its anilide is too long to fit into the capsule. Comparison of the initial rates for the reactions of the two carboxylic acids shows that these rates are quite different, the shorter acid generating products many times faster than the longer one. Indeed, running the reaction in the presence of the shorter anilide shows that the system is autocatalytic as it accelerates its own formation, whereas the longer amide does not. These features are consequences of a feedback loop in the reaction cycle; it leads to chain reaction kinetics, an unprecedented consequence of compartmentalization.

We simulated the rate of DCC release in various situations, and that is shown in the figure. It displays reaction profiles for systems in which one, two or three products can displace the encapsulated DCC. Note that autocatalysis does not occur when only one guest displaces DCC, but with two or three products displacing DCC the sigmoidal profile becomes pronounced. We refer to this kind of behavior as chemical amplification, although it is a form of self-replication as well. Self-replicating systems recognize and select specific molecules for the autocatalytic reaction. In the present case, the DCC is unselective and will react with any carboxylic acid in solution. The other difference in the present case is that product inhibition does not occur because the products do not compete for the reaction sites of the reagents. Instead, they are removed to separate compartments.

The Danishefsky–Gautier Reaction

Another reaction that takes place within the capsule is the reaction of carboxylic acids with isonitriles. While this reaction is more than 100 years old,[13] its recent resurrection by Danishefsky[14] has served peptide synthesis well.[15] The procedure for giving high yield N-formyl

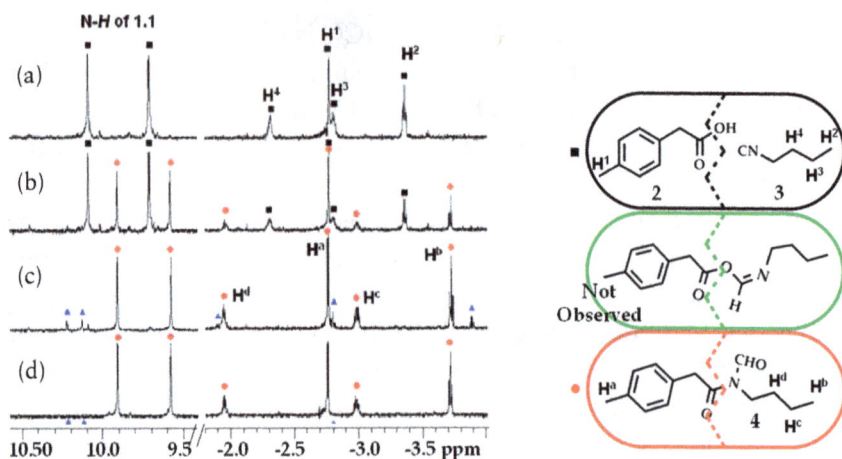

Figure 8.12 Partial NMR spectra for the encapsulated Danishefsky–Gautier reaction; clean conversion from starting materials (trace a) to products (trace d) is observed. [Reprinted from *J Am Chem Soc* **130**, 7810–7811. Copyright 2008, American Chemical Society.]

amides calls for 30 min at 150° in a sealed tube with methylene chloride. The reaction proceeds by addition of the acid to the isonitrile to give an O-acyl-isoamide which undergoes a 1,3-acyl migration process (Mumm rearrangement) to the product (Figure 8.12). Recent work by Houk and Danishefsky[16] indicates that the addition and rearrangement may be concerted processes without the generation of charged intermediates and the strain of a four-membered ring.

We expected that these harsh conditions could be tempered inside a capsule that allows the proper alignment of acids and isonitriles in the first step of the reaction; the second concerted rearrangement would be quite appropriate inside the capsule, where stabilization of charged intermediates would be difficult to imagine. Jun-li Hou and Dariush Ajami chose a combination of reagents that would fill the capsule appropriately, and *p*-tolylacetic acid and *n*-butylisonitrile proved ideal. On mixing with the capsule, an encapsulation complex was formed with one of each reactant inside.[17] Moreover, both acid and isonitrile functions are located near one another in the polar, middle part of the capsule, the reaction proceeds to give the rearrangement product under ambient conditions. When warmed to 40°

the reaction is complete in the course of a day. In the absence of the capsule at comparable (millimolar) concentrations, no reaction could be detected. In fact, outside the capsule in solution at higher (molar) concentrations, *different* products are formed: the formylated amine from the isonitrile and the symmetrical anhydride of the carboxylic acid. Accordingly, the capsule channels the reaction along a different path by isolation of the reagents and intermediates. The reacting centers are well-aligned within the capsule for the formation of transition structures, and the flexible butyl group allows the Mumm rearrangement to take place. The process takes place inside the capsule without observable intermediates.

With a less flexible isonitrile, we were able to observe the initial adduct, the unstable O-acyl-isoamide, by NMR. The same acid coencapsulated with isopropyl isonitrile generates an intermediate species that builds up and then disappears in the spectra (Figure 8.13). From the chemical shift of the isopropyl group, the intermediate is a long structure as both termini are at the ends of the capsule. Apparently,

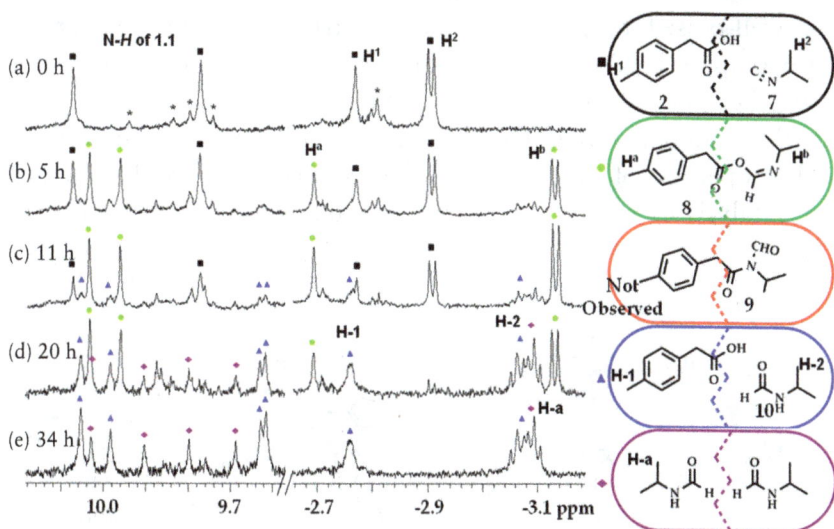

Figure 8.13 Partial NMR spectra for an encapsulated Danishefsky–Gautier reaction reveal an intermediate (trace b) that cannot rearrange within the small space. [Reprinted from *J Am Chem Soc* **130**, 7810–7811. Copyright 2008, American Chemical Society.]

its rearrangement is difficult because the formylated secondary amide is not produced. Instead it is released into solution, where it reacts with carboxylic acid to give the symmetrical anhydride.

Hydration of Isonitriles

The hydration of isonitriles ordinarily requires strongly acidic aqueous conditions, and the hydrophobic nature of most capsules would appear to be an unlikely environment for such a reaction. Nonetheless, the hexameric resorcinarene capsule and the eight H_2O molecules in its hydrogen bonded seam showed effective catalysis of the hydration reaction (Figure 8.14).[18]

Control experiments by the Strukul and Scarso team in Venice established that mere resorcinol Brønsted acidity did not account for the catalysis and competitive inhibition with the dithienylethene isomers shown established that the hydration reaction takes place inside the capsule. Apparently, the protons of the capsule's seam are well-poised to protonate the isonitriles and then deliver an intracomplex H_2O molecule from the concave surface. The water may then be replaced on the convex surface of the capsule by the water-saturated $CDCl_3$ used as solvent. The photophysical properties of the dithienylethenes were employed to place the catalysis of the hydration reaction under some photochemical control.

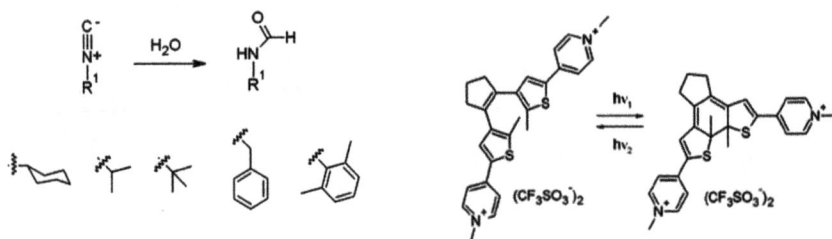

Figure 8.14   *Left*: The hydration reaction of isonitriles is catalyzed by the hexameric resorcinarene. *Right*: The dithienylethene isomers used as competitive inhibitors for the capsule-catalyzed hydration.

Hydrolysis

Following an extensive investigation of the properties of the resorcinarene hexamer (see Chapter 4), Tiefenbacher concluded that the
capsule is a reasonably strong Brønsted acid, with an effective pK_a
of approximately 5.5−6.[19] Typical tertiary amines are protonated
inside, creating cation–π interactions that do much to stabilize the
assemblies. This finding by Tiefenbacher in Munich suggested use of
the capsule as a catalyst for acetal hydrolysis and 1,1-diethoxyethane
(Figure 8.15) was a suitable substrate. In water-saturated $CDCl_3$
under ambient conditions, the capsule showed good conversion of
the acetal to the aldehyde after 1 h, whereas without an added capsule only a slow background reaction occurs. To establish that the
reaction takes place inside, $Bu_4N^+Br^-$ was tested as a competitive
inhibitor; this showed no detectable hydrolysis. Surprisingly, larger
substrates indicated reduced rates with the capsule, even though modeling showed that they could be accommodated inside.

Rearrangements Inside Capsules

Inhibition

The Danishefsky–Gautier reaction inside the capsule gives product
molecules that are shorter than starting material molecules, and we
were curious to see if the limited space could arrest reactions in the
capsule that gave product molecules that were longer than the space
allowed. Or would the reaction process force the rupture of the capsule? As seen in Chapter 3, the photophysics of azobenzene isomerization *does* break open capsules, but no examples of simple chemical
processes were known to do this.

Aaron Sather decided to study the decomposition of N-nitrosoamides and their generation of esters and related compounds
(Figure 8.16).[20] This reaction is also believed to proceed through
a Mumm-like rearrangement, and like the 1,3-acyl shift, it may
involve a four-membered charge-separated intermediate, or it may
rearrange concertedly. In either event, the rearrangement product is

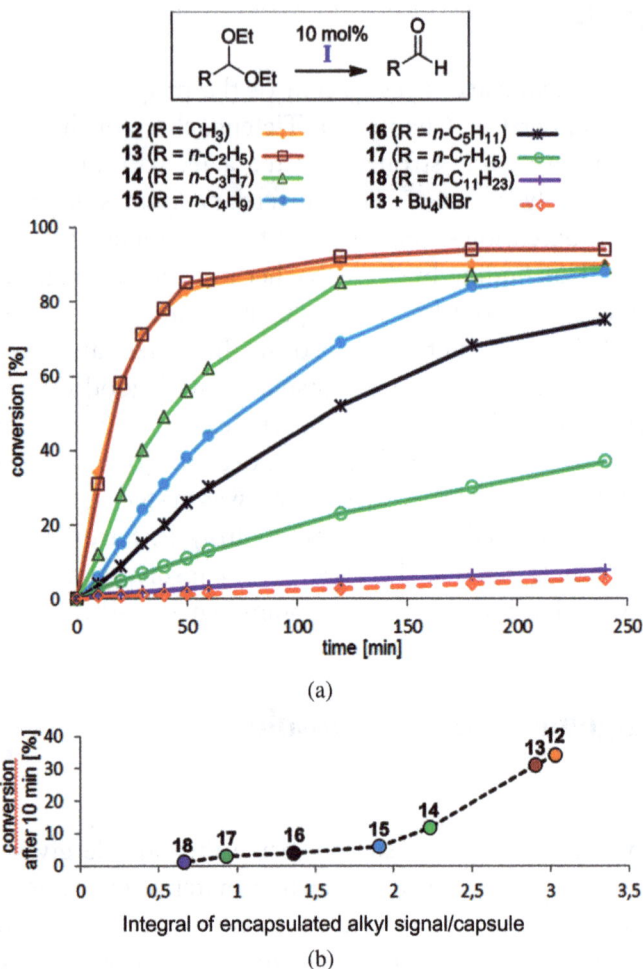

Figure 8.15 *Top*: Product evolution curves for the hydration of acetals in the hexameric resorcinarene capsule. *Bottom*: The same reaction in the absence of a capsule in CDCl₃ solvent. [Reprinted with permission from *J Am Chem Soc* **135**, 16213–16219. Copyright 2013, American Chemical Society.]

longer as an N–O unit is covalently inserted into the center of the molecule.

Modeling shows that *p*-toluic amide as its N-nitroso-N-butyl derivative fits quite nicely into the cylindrical capsule (Figure 8.16). Typically, such aromatic derivatives decompose at reasonable rates

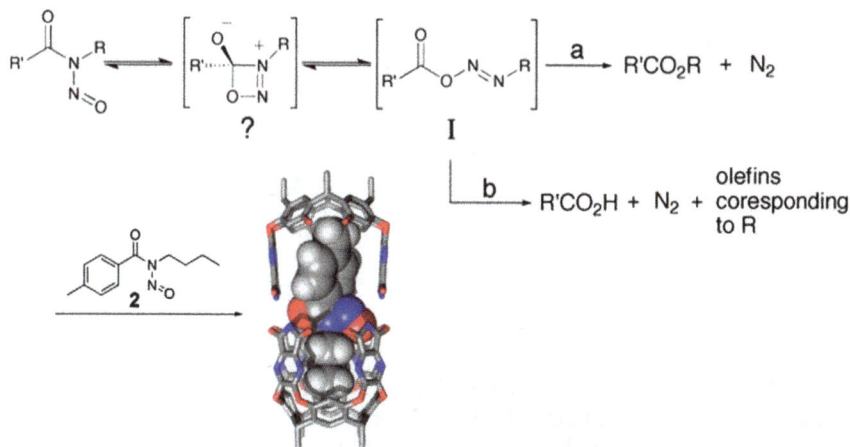

Figure 8.16 *Top*: The decomposition of N-nitroso-amides generates esters and related olefinic compounds. *Bottom*: A butyl amide derivative modeled in the cylindrical capsule. [Reprinted from *Tetrahedron Lett* 52(17), Reactivity of N-nitrosoamides in confined spaces, pp. 2100–2103. Copyright 2011, with permission from Elsevier.]

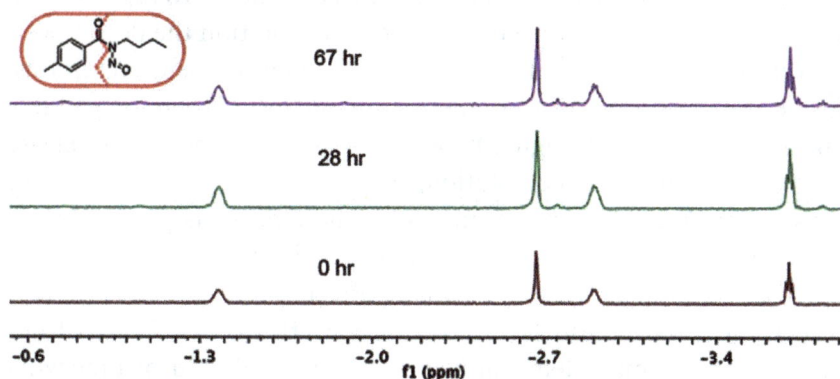

Figure 8.17 Partial NMR spectra reveal that the encapsulated N-nitroso compound was stable indefinitely to mild heating. [Reprinted from *Tetrahedron Lett* 52(17), Reactivity of N-nitrosoamides in confined spaces, pp. 2100–2103. Copyright 2011, with permission from Elsevier.]

on gentle warming. However, the encapsulated compound was found to remain stable indefinitely on mild heating (Figure 8.17).

A longer molecule, the N-hexyl derivative, proved a reluctant guest. However, its rearrangement product, the hexyl ester, was a

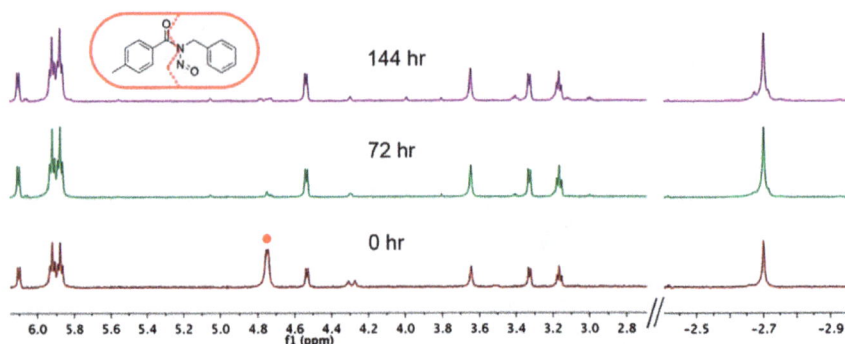

Figure 8.18 Partial NMR spectra reveal that the encapsulated N-benzyl, N-nitroso toluamide was also stable to mild heating. [Reprinted from *Tetrahedron Lett* 52(17), Reactivity of N-nitrosoamides in confined spaces, pp. 2100–2103. Copyright 2011, with permission from Elsevier.]

good guest. We showed that reaction of the hexyl compound takes place *outside* the capsule, followed by guest exchange. We also studied the N-benzyl-N-nitroso derivatives, which are known to rapidly form benzyl cations on rearrangement. Indeed, in solution the decomposition occurs with a half-life of three days under ambient conditions. Inside the capsule, however, the compound was indefinitely stable and completely intact (Figure 8.18). On addition of methanol, release of the compound to the solution occurred and the rearrangement took place. Accordingly, the snug fit of these reagents prevents their rearrangement in the capsule and stabilizes them.

We encountered a more dramatic effect involving the inhibition of peroxide decomposition. In this case, Steffi Körner, Dmitri Rudkevich and Thomas Heinz showed that encapsulation of dibenzoyl peroxide protects it from both external reagents (acids and bases) and internal decomposition. This peroxide is stable for more than three days at 70°C. Outside the capsule, decomposition is complete in 3 h. For slow release of the peroxide, *p*-[N-(*p*-tolyl)]toluamide can be added and the release half-life takes 3–5 days. The bulkier *p*-[N-methyl-N-(*p*-tolyl)]toluamide takes weeks to release the peroxide, but a small amount of DMF will release the peroxide within seconds. The DMF ruptures the hydrogen bond seam that holds the capsule together (Figure 8.19). Triphenyl phosphine has no effect on the encapsulated

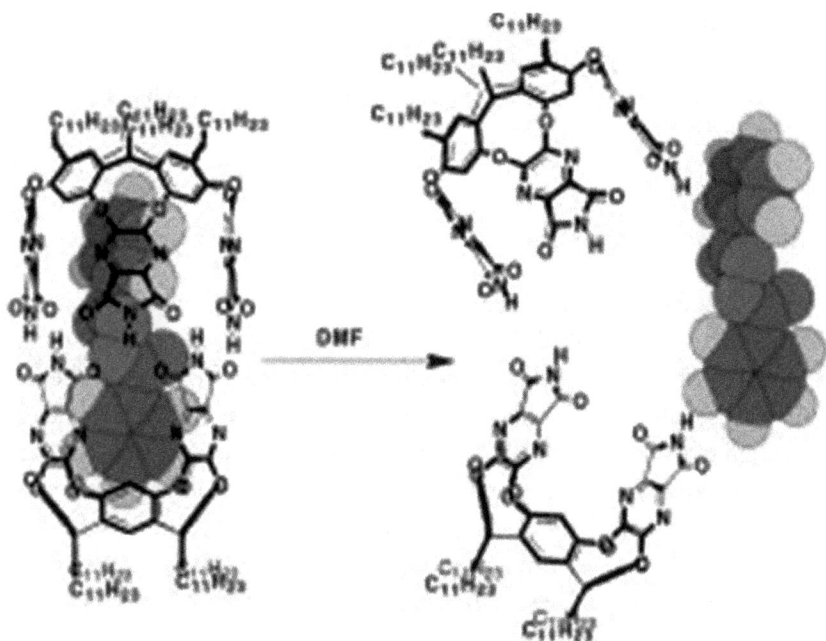

Figure 8.19   The release of dibenzoyl peroxide by DMF takes place within seconds as the hydrogen bond seam that holds the capsule together is ruptured.

peroxide but, on release, rapid reaction occurs, as shown by ^{31}P NMR spectroscopy. Likewise, a solution of 1,5-diphenyl carbazide, a colorimetric reagent for peroxide determination, showed no reaction with peroxide inside the capsule. However, on release of the peroxide after several days, a full colorimetric response takes place.

In short, the capsule *protects* dibenzoyl peroxide. It is not known whether dissociation occurs on heating inside to give carboxyl radicals that have no choice but to recombine. This could be tested by ^{18}O labeling. It is also likely that the typical mode of decomposition — decarboxylation — is forbidden in the capsule. Decarboxylation would give an already well-filled capsule (two phenylcarboxyl radicals) a third species (CO_2, a phenyl radical and a phenylcarboxyl radical) to accommodate. This is unfavorable, as gases require even more space inside capsules. Whatever the cause, perhaps an application can be found where a reactive species could be stored inside the capsule for a long time, then released whenever it is needed into the bulk solution. A spectacular case of stablilizing white phosphorous in a different type of capsule was recently described by Rissanen and Nitschke.[21]

Catalysts in Capsules

The first encapsulated Au catalyst (Figure 8.20) in organic solution was reported by Strukul, Reek, Scarso, *et al.*[22] Its application in the hydration of terminal alkynes yielded products of different chemo- and regioselectivities in the resorcinarene hexamer than was typically observed in bulk solution. The limited space inside the capsule favored intramolecular reactions attributed to the unusual folding of the substrates. Specifically, the 1,2-dihydronaphthalene product is formed by intramolecular rearrangement and is usually observed only in the absence of water. The resorcinarene capsule's interior is hydrophobic and is expected to slow the substrate's reaction with water.[23] However, this result is different than the same capsule's catalysis of acetal hydration through the delivery of water from the seam of hydrogen bonds. Solvation of the catalyst in the hexameric host alters its activity as well as selectivity in a way reminiscent of catalysis at the active site of enzymes.

Extensive studies of Au catalysts encapsulated in resorcinarene hexamers have been reported by Ballester and Echvarren.[24] These include new methods of encapsulation involving an excess of water,

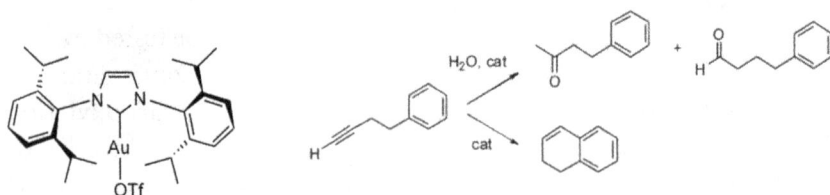

Figure 8.20 *Left*: A gold catalyst used in the resorcinarene hexamer. *Right*: The limited space inside the capsule favored products with different regioselectivities than seen in bulk solution.

and means by which different encapsulated species can be prepared from the same gold(I) precursor.

While enzymes are well known to stabilize reactive intermediates and high energy transition structures, synthetic capsules can do so too. The grandfather of all these is Cram's taming of cyclobutadiene inside a covalent capsule — a carcerand.[25] Although we[26] and others[27] had generated indirect evidence for the fleeting existence of cyclobutadiene, seeing is believing and Cram's direct observation of this remarkable structure stands as a landmark in physical organic chemistry. Earlier, Mayer had shown that with the proper tert-butyl substitutions, the cyclobutadiene[28] and its tetrahedrane[29] valence isomer could be isolated and characterized. Since then, the direct observation of a number of reactive intermediates has been performed with both covalent and hydrogen-bonded structures. Synthetic enzyme mimics abound.[30]

Reaction Selectivity

Here, we examine some ring–chain tautomerism reactions in which the capsule amplifies the unstable molecules and, in some cases, reveals molecules that have no existence outside of the capsule. Consider the equilibrium between the Schiff base and the heterocycle of Figure 8.21. In typical organic solvents, the open chain form is favored in concentration by about 10-fold, although both are present. However, Tetsuo Iwasawa and Enrique Mann exposed this equilibrium to the capsule in mesitylene, and saw that the ring compound was present at 90% in the capsule, while the open chain was present at a mere 10%.[31] In the capsule the relative stabilities were completely reversed. This result is somewhat general, as the more rigid compounds are more comfortable inside the capsule for entropic reasons.

The second example involved the related dimethylamino compounds. With these, the heterocycle is already favored in solution, but inside the capsule that isomer is almost exclusively present. There is not an obvious pathway for interconversion inside the capsule. This would require charged intermediates and clumsy proton transfers inside, and gives us reason to believe that the capsule captures the

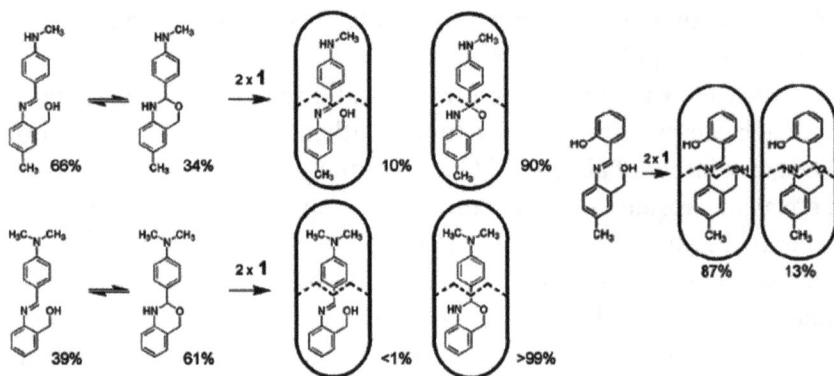

Figure 8.21   Some ring–chain tautomerism reactions of Schiff bases in the cylindrical capsule. The capsule amplifies the more rigid ring isomers even when these are not favored in bulk solution. [Reprinted with permission from *J Am Chem Soc* **128**, 9308–9309. Copyright 2006, American Chemical Society.]

isomers and siphons them off to give the new equilibrium concentrations inside. The third system, involving the salicylaldehyde derivatives, gave results that were even more impressive.

Salicylaldehyde Schiff bases have an advantage in reactivity, because the *o*-hydroxyl group could provide intramolecular acid catalysis for the breakdown or formation of the heterocyclic compound. This possibility warranted a close look. No ring isomers were detectable in mesitylene solution for the corresponding Schiff base, yet in the capsule somewhat more than 10% of this compound can be observed — again preferentially stabilized (Figure 8.21). It is quite likely that the reaction takes place inside the capsule through intamolecular acid catalysis by the phenolic hydroxyl group. An equivalent but more dramatic interpretation is that the heterocyclic compound *does not exist outside the capsule*.

Studies at different temperatures showed that the equilibrium constant for the salicylaldehyde case remained the same over the span of nearly 100°. In other words, ΔH is approximately 0, and entropy determines the free energy, ΔG. It should be possible to stabilize other tetrahedral intermediates, such as hemiaminals or hemiacetals, inside this capsule, where the high molar concentrations may reveal otherwise invisible intermediates. Raymond and Bergman[32] have seen such fragile species inside their charged capsules, held together by ligand–metal interactions in water. Fujita has also given many examples in related capsules,[33] and we have done so too with open-ended cavitand container molecules.[34]

Epilogue

Before we end, let us state the following: There is nothing wrong with writing about molecular behavior. It is not that we anthropomorphize molecules — it is about thinking of why something partitions between two different environments. The molecules are thinking: "What am I doing here, in this uncomfortable and very small space?" Answer: "I'm just trying to get out of the solvent. The solvent out there irritates my surface, and my presence annoys the solvent — it has to organize itself around me. What can I do? How do I get out of here? I'll just separate into my own phase, but I still have some exposure. It's a small, hydrophobic space; let me try folding in half. Ahh, that's better. Under these conditions that's the best I can do, and the solvent is relieved, too." There is no doubt that enzymes behave this way. Their very folding involves this monologue. The small molecules get out of the water and fit it into a very small space even though it has to compress itself. Who knew that an alkane would be the simplest foldamer?

What has been learned from molecular behavior in small spaces? For more than a century physical organic chemistry was practiced in dilute solution. The reagents were surrounded by a sea of solvent, an environment where encounters were random, short-lived, and controlled by diffusion under ambient conditions. The present monograph describes the behavior in limited spaces defined by mechanical barriers where encounters are long-lived, intense and intimate. The container may be regarded as an almost permanent solvent cage, fixed by synthesis. We have seen that subtle changes in the capsule or guest structure give discontinuous differences in the behavior of the guest molecules in the limited spaces. The difficulty in predicting which small geometrical changes in one of the building blocks affect the outcome of multicomponent assemblies has been encountered by others.[35,36] Finally, we note that unlike the covalent carcerands or cryptophanes, with well-defined structures, some of these unanticipated assemblies do not exist on their own; they appear only when a fortunate combination of host and guest is present. The instructions are written in their curvature, hydrogen-bonding patterns and complementarity, but filling their small spaces properly is the universal, final instruction for the assembly to emerge.

In encapsulation complexes, as in architecture, the space that is created by a structure limits what goes on inside. Unlike architecture, the capsules cannot be created or assemble without anything inside. We have never encountered an empty capsule of the hydrogen-bonded variety. A recognition event precedes almost all bimolecular reactions, and physical organic chemists concerned with molecular recognition used organic solvents, where most synthetic reactions are performed, rather than water. But molecular recognition in water is a challenge that is one of the outstanding problems of the day. Discoveries like the enthalpic hydrophobic effect,[37] the cation–π interactions[38] and the 55% solution[39] all grew out of physical organic chemistry using synthetic receptors that largely surrounded their targets in either water or other solvents. These explorations have been satisfyingly enhanced by hydrogen-bonded capsules.

Fudan University, Shanghai and
The Scripps Research Institute, La Jolla, 2015

References

1. Kang J, Rebek Jr J. (1997) Acceleration of a Diels–Alder reaction by a self-assembled molecular capsule. *Nature* **385**, 50–52.
2. Kang J, Hilmersson G, Santamaria J, Rebek Jr J. (1998) Diels–Alder reactions through reversible encapsulation. *J Am Chem Soc* **120**, 3650–3656.
3. Kang J, Santamaria J, Hilmersson G, Rebek Jr J. (1998) Self-assembled molecular capsule catalyzes a Diels–Alder reaction. *J Am Chem Soc* **120**, 7389–7390.
4. Yoshizawa M, Tamura M, Fujita M. (2006) Diels–Alder in aqueous molecular hosts: unusual regioselectivity and efficient catalysis. *Science* **312**, 251–254.

5. Shimizu S, Usui A, Sugai M, *et al.* (2013) Hexameric capsule of a resorcinarene bearing fluorous feet as a self-assembled nanoreactor: a Diels–Alder reaction in a fluorous biphasic system. *Eur J Org Chem* 4734–4737.

6. Shimizu S, Kiuchi T, Pan N. (2007) A "Teflon-footed" resorcinarene: a hexameric capsule in fluorous solvents and fluorophobic effects on molecular encapsulation. *Angew Chem Int Ed* 46, 6442–6445.

7. Percec V, Johansson G, Ungar G, Zhou J. (1996) Fluorophobic effect induces the self-assembly of semifluorinated tapered monodendrons containing crown ethers into supramolecular columnar dendrimers which exhibit a homeotropic hexagonal columnar liquid crystalline phase. *J Am Chem Soc* 118, 9855–9866.

8. Kolb HC, Finn MG, Sharpless KB. (2001) Click chemistry: diverse chemical function from a few good reactions. *Angew Chem Int Ed* 40, 2004–2021.

9. Rostovtsev VV, Green LG, Fokin VV, Sharpless KB. (2002) A stepwise Huisgen cycloaddition process: copper(I)-catalyzed regioselective "ligation" of azides and terminal alkynes. *Angew Chem Int Ed* 41, 2596–2599.

10. Chen J, Rebek Jr. J. (2002) Selectivity in an encapsulated cycloaddition reaction. *Org Lett* 4, 327–329.

11. Chen J, Körner S, Craig SL, *et al.* (2002) Amplification by compartmentalization. *Nature* 415, 385–386.

12. Chen J, Körner S, Craig SL, *et al.* (2002) Chemical amplification with encapsulated reagents. *Proc Natl Acad Sci USA* 99, 2593–2596.

13. Gautier A. (1896) Ueber die einwirkung der Sauern auf die carbylamine. *Liebigs Ann* 151, 240—243.

14. Li X, Danishefsky SJ. (2008) New chemistry with old functional groups: on the reaction of isonitriles with carboxylic acids — a route to various amide types. *J Am Chem Soc* 130, 5446–5448.

15. Wilson MR, Stockdill JL, Wu X, *et al.* (2012) A fascinating journey into history: exploration of the world of isonitriles en route to complex amides. *Angew Chem Int Ed* 51, 2834–2848.

16. Jones GO, Li X, Hayden AE, *et al.* (2008) The coupling of isonitriles and carboxylic acids occurring by sequential concerted rearrangement mechanisms. *Org Lett* 10, 4093.

17. Hou JL, Ajami D, Rebek Jr J. (2008) Reaction of carboxylic acids and isonitriles in small spaces. *J Am Chem Soc* 130, 7810–7811.

18. Bianchini G, La Sorella G, Canever N, *et al.* (2013) Efficient isonitrile hydration through encapsulation within a hexameric self-assembled capsule and selective inhibition by a photo-controllable competitive guest. *Chem Commun* 49, 5322–5324.

19. Zhang Q, Tiefenbacher K. (2013) Hexameric resorcinarene capsule is a Brønsted acid: investigation and application to synthesis and catalysis. *J Am Chem Soc* 135, 16213–16219.

20. Sather AC, Berryman OB, Ajami D, Rebek Jr J. (2011) Reactivity of N-nitrosoamides in confined spaces. *Tetrahedron Lett* 52, 2100–2103.

21. Mal P, Breiner B, Rissanen K, Nitschke JR. (2009) White phosphorus is air-stable within a self-assembled tetrahedral capsule. *Science* 324, 1697–1699.

22. Cavarzan A, Scarso A, Sgarbossa P, *et al.* (2011) Supramolecular control on chemo- and regioselectivity via encapsulation of (NHC)-Au catalyst within a hexameric self-assembled host. *J Am Chem Soc* **133**, 2848–2851.

23. Cavarzan A, Reek JNH, Trentin F, *et al.* (2013) Substrate selectivity in the alkyne hydration mediated by NHC-Au(I) controlled by encapsulation of the catalyst within a hydrogen bonded hexameric host. *Catal Sci Tech* **3**, 2898–2901.

24. Adriaenssens L, Escribano-Cuesta A, Homs A, *et al.* (2013) Encapsulation studies of cationic gold complexes within a self-assembled hexameric resorcin[4]arene capsule. *Eur J Org Chem* 1494–1500.

25. Cram DJ, Tanner ME, Thomas R. (1991) The taming of cyclobutadiene. *Angew Chem Int Ed* **30**, 1024–1027.

26. Rebek Jr J, Gavina F. (1974) The three-phase test for reactive intermediates cyclobutadiene. *J Am Chem Soc* **96**, 7112–7114.

27. Grubbs RH, Grey RA. (1973) Cyclobutadiene as an intermediate in the oxidative decomposition of cyclobutadienyliron tricarbonyl. *J Am Chem Soc* **95**, 5765–5767.

28. Irngartinger H, Riegler N, Malsch KD, *et al.* (1980) Structure of tetra-tert-butylcyclobutadiene. *Angew Chem Int Ed Eng* **19**, 211–212.

29. Maier G, Pfriem S, Schäfer U, Matusch R. (1978) Tetra-tert-butyltetrahedrane. *Angew Chem Int Ed Eng* **17**, 520–521.

30. See, for example: Raynal M, Ballester P, Vidal-Ferrana A, van Leeuwen PWNM. (2014) Supramolecular catalysis. Part 2: Artificial enzyme mimics. *Chem Soc Rev* **43**, 1734–1787.

31. Iwasawa T, Mann E, Rebek Jr J. (2006) A reversible reaction inside a self-assembled capsule. *J Am Chem Soc* **128**, 9308–9309.

32. Pluth MD, Fiedler D, Mugridge JS, *et al.* (2009) Molecular recognition and self-assembly special feature: encapsulation and characterization of proton-bound amine homodimers in a water-soluble, self-assembled supramolecular host. *Proc Natl Acad Sci USA* **106**, 10438–10443.

33. Fujita M, Umemoto K, Yoshizawa M, *et al.* (2001) Molecular paneling via coordination. *Chem Commun* **6**, 509–518.

34. Iwasawa T, Hooley RJ, Rebek Jr J. (2007) Isolation and observation of unstable intermediates in carbonyl addition reactions. *Science* **317**, 493–496

35. Sun QF, Iwasa J, Ogawa D, *et al.* (2010) Self-assembled M24L48 polyhedra and their sharp structural switch upon subtle ligand variation. *Science* **328**, 1144–1147.

36. Lin Z, Emge TJ, Warmuth R. (2011) Multicomponent assembly of cavitand-based polyacylhydrazone nanocapsules. *Chem Eur J* **17**, 9395–9405.

37. Ferguson SB, Seward EM, Diederich F, *et al.* (1988) Strong enthalpically driven complexation of neutral benzene guests in aqueous solution. *J Org Chem* **53**, 5593–5595.

38. Shepodd TJ, Petti MA, Dougherty DA. (1988) Molecular recognition in aqueous media: donor-acceptor and ion-dipole interactions produce tight binding for highly soluble guests. *J Am Chem Soc* **110**, 1983–1985.

39. Mecozzi S, Rebek Jr J. (1998) The 55% Solution: a formula for molecular recognition in the liquid state, *Chem Eur J* **4**, 1016–1022.

Index